Javelin – From The Cockpit

JAVELIN – FROM THE COCKPIT

Britain's First Delta Wing Fighter

by

Peter Caygill

Pen & Sword
AVIATION

First published in Great Britain in 2011 by
Pen & Sword Aviation
an imprint of
Pen & Sword Books Ltd
47 Church Street
Barnsley
South Yorkshire
S70 2AS

Copyright © Peter Caygill 2011

ISBN: 978-1-84884-656-2

Typeset in 12/13.5pt Palatino by
Concept, Huddersfield, West Yorkshire

Printed and bound in England by
CPI Group (UK) Ltd, Croydon, CRO 4YY

Pen & Sword Books Ltd incorporates the Imprints of Pen & Sword
Aviation, Pen & Sword Family History, Pen & Sword Maritime, Pen &
Sword Military, Pen & Sword Discovery, Wharncliffe Local History,
Wharncliffe True Crime, Wharncliffe Transport, Pen & Sword Select,
Pen & Sword Military Classics, Leo Cooper, The Praetorian Press,
Remember When, Seaforth Publishing and Frontline Publishing.

For a complete list of Pen & Sword titles please contact
PEN & SWORD BOOKS LIMITED
47 Church Street, Barnsley, South Yorkshire, S70 2AS, England
E-mail: enquiries@pen-and-sword.co.uk
Website: www.pen-and-sword.co.uk

Contents

Acknowledgements

Much of the research for this book was carried out at the National Archives at Kew, in particular flight trials reports on the various marks of Javelin as conducted by A&AEE Boscombe Down and AWDS at West Raynham. I also looked at relevant RAF squadron records and accident reports and would like to thank the staff at TNA not only for their help, but their extremely efficient service in delivering the documents that provided much of the information for this work.

This book could not have been written without the enthusiastic support of a number of former Javelin pilots and navigators and I am deeply indebted to them. Their testimonies provide a fascinating insight into the Javelin and RAF squadron life in the 1950s and 1960s. In alphabetical order they are as follows : Paul Burns (60 and 29 Squadrons), Fred Butcher (23 and 60 Squadrons), Noel Davies (AWDS), Colin Edwards (46 and 60 Squadrons), Howard Fitzer (46 and 23 Squadrons), Paddy Harbison (29 Squadron), Don Headley (64 Squadron), Paul Hodgson (89 Squadron), the late Ivan Logan (23 Squadron), Mike Miller (23 and 60 Squadrons), Hank Prosser (41 Squadron), Jim Sawyer (33 Squadron), Gerry Shipley (23 and 60 Squadrons), Bruce Spurr (228 OCU), Pete Thorn (23 and 29 Squadrons), Tony Warner (46 Squadron) and Guy Woods (151 and 25 Squadrons).

I would also like to thank Bob Hillard who was a ferry pilot with 187 Squadron and flew the very first Javelin FAW.1 to enter RAF squadron service, and Peter Hicks who was involved in a mid-air collision with a Javelin when flying a Hawker Hunter of the Central Flying School.

The photographs in this book came from a variety of sources and in particular I would like to thank Philip Jarrett, Howard Fitzer, Mike Miller and Colin Edwards for their help in supplying suitable images.

Finally I would like to thank Peter Coles and the production staff at Pen and Sword for their help and assistance with this project.

Peter Caygill
Darlington

Introduction

During its time in RAF service the Gloster Javelin acquired a few nicknames, most of them uncomplimentary. The inspiration for these names is fairly obvious, 'Flatiron' was a commentary on the shape of its delta wing and the term 'Dragmaster' came about as a result of its size and proportion, in particular its thick wing. The Javelin was an imposing aircraft especially for those RAF pilots who had made the transition from day fighters to all-weather fighters and had swapped the beautiful lines and crisp handling of the Hawker Hunter for something that on first acquaintance looked more like a bomber. For pilots and navigators who were already flying in night-fighter squadrons the Javelin was warmly welcomed as a being huge advance in performance and capability when compared to the Meteors and Venoms that they had been used to.

The Javelin had a troubled gestation and did not enter RAF squadron service until 1956 which was roughly ten years after the initial requirement had been formulated. It was effectively a 'bomber-destroyer' and was designed to operate against Soviet bombers armed with nuclear weapons. In the event of a nuclear conflict it was assumed that aircraft such as the Tupolev Tu-16 Badger would be deployed against the UK in mass raids and it was the job of the Javelin crews to intercept and destroy these aircraft as far from the UK coast as possible. The early marks of Javelin were armed with four 30 mm Aden cannon, however later marks introduced the Firestreak air-to-air missile which provided a quantum leap in operational capability. Although the early Javelins were considered by some to be under-powered, the later Javelins also had more powerful Sapphire engines which boosted performance and allowed interceptions to be made up to 50,000 ft

The Javelin was the RAF's principal night/all-weather fighter until the advent of the English Electric Lightning, but once its 'Cold War' duties were over it provided a valuable contribution in various trouble-spots throughout the world, in particular the Indonesian-Malaysian Confrontation in the Far East. It also provided a much-needed British presence in the Mediterranean with regular detachments to Cyprus. From March 1963 No. 29 Squadron was based at Nicosia before moving to Akrotiri and it was from here that ten aircraft detached to Zambia to provide air-support following Rhodesia's Unilateral Declaration of Independence in November 1965.

Although it did not inspire the affection of the Hunter, the Javelin was well-liked in service and did everything that was asked of it. Over the years it has become something of an 'unsung hero' and this book attempts to redress the balance. The story of the Javelin is told from the crew's point of view, both from information derived from flight trials reports by the test establishments at Boscombe Down and West Raynham and numerous first-hand accounts by pilots and navigators. Their testimonies show that the Javelin was an able performer and was developed into a first-rate weapons system.

CHAPTER 1

Design and Development

The Gloster GA.5 (later to be called Javelin) was flown for the first time on 26 November 1951 by Gloster Chief Test Pilot S/L W.A. 'Bill' Waterton AFC, the flight marking the end of a protracted gestation which can be traced back to the end of the Second World War. With the fall of Germany in 1945 there was a race between the victorious Allied nations to discover its secrets, in which particular emphasis was placed on acquiring information relating to Germany's advanced aircraft projects. The discovery of radical new shapes for aircraft and associated aerodynamic data was to have a profound effect on the design of what would become the Javelin.

In the period immediately after the Second World War the night skies over Britain were patrolled by night-fighter versions of the de Havilland Mosquito which was equipped with the American-manufactured centimetric-wave AI.10 radar. Although one of the classic designs of the war, the pace of aeronautical development in the immediate post war years was such that the Mosquito quickly became obsolescent. Existing long-range bombers such as the Boeing B-29 Superfortress (soon to be cloned by the Soviet Union as the Tupolev Tu-4) were capable of cruising at speeds of around 300 mph at altitudes of 35,000 ft, and in the near future there was the prospect of jet bombers entering service with much greater performance. In the face of such a potential threat it was clear that Britain's air defences, both day and night, needed a radical transformation and a number of specifications were issued by the Air Ministry.

In January 1947 Specifications F.43/46 for a single-seat interceptor and F.44/46 for a two-seat night-fighter were issued. The requirements of the latter included a maximum speed of not less than 525 knots with a service ceiling of 45,000 ft, this altitude to be attained in no more than ten minutes after engine start. As Gloster Aircraft had pioneered jet-powered flight in the UK with the single-engine experimental E.28/39 and the twin-engine Meteor day fighter, the company was ideally placed to submit proposals. Their first response carried the project number P.228 and the basic layout showed similarities to the Meteor with two Rolls-Royce AJ.65 (later Avon) jet engines mounted in the wings. By April 1947 a rather different aircraft (the P.240) was being proposed which had wings of reduced aspect ratio and more leading edge sweep. The engines had by now been moved inboard to the fuselage sides and the tail plane was placed higher on the fin.

Gloster also put forward schemes for a single-seat interceptor and one of these, the P.234 which pre-dated the P.240 by two months, was the first to show a delta wing. The use of a wing of this shape stemmed from the pioneering work carried out by Alexander M. Lippisch in Germany who had flown delta-winged gliders and powered aircraft in the 1930s. The delta wing offered a number of advantages over swept wings, not least of which was that the extended root chord allowed the wing to be built thicker whilst keeping thickness/chord ratio acceptably low to delay the onset of compressibility problems. It also allowed the wing to be built lighter and reduced any tendency towards aileron reversal due to wing twist when high loadings were imposed. The delta wing of the P.234 was not its only radical

4

feature as it also had a delta butterfly tail. A further delta wing proposal was the P.238 of March 1947 that was to be powered by two Metropolitan Vickers F9 jet engines (shortly to become the Armstrong Siddeley Sapphire). It featured a horizontal delta tail plane mounted just below the top of the fin.

In February 1948 F.43/46 and F.44/46 were superseded by revised specifications F.3/48 and F.4/48 respectively. The Gloster proposal relating to F.4/48 was the P.272 which bore a marked resemblance to what became the Javelin. The aircraft was to be powered by two Sapphire 2 engines of 9,000 lbs thrust but there were notable differences to the Javelin, in particular the use of rotating wingtips instead of ailerons for lateral control and a slab tail plane. The design continued to evolve and eventually settled on a design showing a wing with an area of 900 sq.ft, with power provided by two Sapphire engines of 7,500 lbs thrust. After one last single-seat day fighter proposal (P.275) Gloster decided to concentrate on the two-seat night/all-weather fighter and on 13 April 1949 they were rewarded with an order for four prototypes. At the same time de Havilland was awarded a contract for its F.4/48 proposal, the DH.110. Unfortunately the Gloster F.4/48 suffered from political meddling as the order was reduced to two later in the year as an economy measure, a decision that seriously delayed the development programme (the order was subsequently increased to six in March 1951!).

The delta planform was adopted by the UK more than any other nation but although the potential benefits were known, there was little knowledge of the likely handling characteristics. Research aircraft such as the Boulton Paul P.111 and P.120 were ordered, but delays meant that the information that these aircraft provided was too late to influence the design of the F.4/48 which by now had acquired the Gloster designation GA.5. The prototype GA.5 (serial number WD804) slowly began to take shape at the Gloster factory at Bentham and in July 1951 it was taken to bits again and moved to the airfield at Moreton Valence. Here it was reassembled and prepared for the testing work it was about to undertake.

The first flight of the GA.5 was not without incident as buffeting and vibration was apparent, this appearing to emanate from the rear of the aircraft so that speed had to be kept below 200 kts IAS. On landing the problem was immediately obvious however as oil staining on the rudder showed that the airflow in the area was disturbed by inter action with the jet efflux. The buffeting was eventually overcome by a progressive lengthening of the fairing around the jet-pipes which ultimately resulted in the familiar 'pen-nib' shape. Although the GA.5 was easy enough to fly and showed great promise it could bite if pilots were complacent. For instance when the flaps were lowered on the approach there was significant pitch up which required the stick to be moved sharply forward together with a lot of tail plane trim. The prototype GA.5 could also be a tiring aircraft to fly, especially during a low level aerobatic sequence. In his autobiography *The Quick and the Dead* (Frederick Muller 1956), Bill Waterton recalls the effort that was needed and the fact that he had to resort to tail plane trim at times and keep both hands on the control column to manoeuvre the aircraft while keeping speed relatively low [the GA.5 was a big aircraft and its all-up weight was twice as much as the prototype Hawker Hunter single-seat fighter. At the time its controls were also far from the standard of later production aircraft].

For its early testing, WD804 was fitted with a fairing on top of the tailplane for an anti-spin parachute but this had been removed by April 1952. Unusually for a first prototype there were very few delays to the test schedule due to unserviceability and the flight envelope was gradually extended up to an altitude of 43,000 ft and a speed of Mach 0.91. By the end of June 1952 WD804 had completed ninety-eight test flights but was still far from being the finished article as continual adjustments to the 'feel' of the power-assisted controls needed to be tested in the air to see if the expected improvement in the aircraft's high speed handling characteristics had materialised. On 29 June Bill Waterton took off for one more test flight but was to experience a severe case of elevator flutter during a high speed run at low level over Oxfordshire. Many years later he recalled the day's events in a letter to Len Thorne, a former colleague at the Air Fighting Development Unit at Wittering

The Javelin was to be shown off together with all other new fighter prototypes at CFE, West Raynham, so on the Sunday afternoon before it was due to go I took off from Moreton Valence, the Company's aerodrome just south of Gloucester to check it over. I went towards Oxford, then turned west near Brize Norton to return to base at 3,000 ft. As I was still accelerating past 560 knots IAS there occurred a small, sharp bump followed by extreme, medium amplitude up-and-down vibration, blurred sight, this being accompanied by a buzzing noise like an angry bumble bee. Shortly afterwards (about 2½ seconds) there was a loud crack followed by another which brought a complete cessation of vibration and noise. Both elevators had broken off at the hinges. The corrected speed of the aircraft when this happened was just beyond 680 mph.

I advanced the throttles fully and the nose came up into a slight climb during which time I found that there was slow trim by tail plane movement via the trim wheel and screw jack. I was then able to throttle back and climb to 10,000 ft thinking to bale out over the Bristol Channel where the aircraft would not strike land and cause any damage. I then had second thoughts as the Gloster boffins would never believe me when I reported what had happened, assuming I was still around to be able to tell them. I found I could fly it down to 160 knots without flaps which would have pitched it upwards, perhaps beyond the trim-out capacity. The long east–west runway at Boscombe Down with a clear approach was a good choice. I made a good touchdown at 160 knots, but gusts of wind and runway irregularities tossed the aircraft up (people said it would have cleared a hangar in height) and after 2–3 of these gigantic bounces the port undercarriage leg was driven up from the spar and through the port fuel tank which exploded. The aircraft then pivoted round the port wing tip, skidded off the runway and stopped in a heap, the starboard under-carriage leg having also now collapsed.

So far, not so good, but I was still alive, although jarred and bruised. At least the aircraft stopped upright, although by now there was some 300 gallons of fuel going up in flames and the port side of the canopy was melting. I had to get out but the canopy appeared to be jammed and the electrics had gone. I was then aware of an explosion as the starboard wing tank went up in flames. Two young RAF officers had been running towards my blazing wreck but when the starboard tank pooped off they turned about and all I could see were two arses pumping up and down and rapidly disappearing behind the watch tower. I heard myself saying aloud 'You poor pair of pricks – you're leaving me here to cook!'

As I was trapped I shut off the oxygen, thinking not to encourage fires then realised the future didn't appear too rosy. The Martin Baker ejection seat was propelled by the shell of a 3 in artillery piece and the heat could cook it off like an overheated breech of a rifle could set off a round. Also at that time Martin Baker seats had no Perspex cutters atop them like the later models did. I pushed and bashed away at the canopy overhead; it opened enough to get my fingers into the gap and with much effort the gap became large enough for me to extract myself from the aeroplane. The flames on the starboard side were now much reduced and I was able to get at the recorder panel ahead of the cockpit and retrieve the early type 'black box'. I was put to bed by the station quack 'doing his stuff' but was up again after an hour awaiting a car from the firm. For once my girlfriend (later wife) was actually pleased to see me! The next day I flew a Dragon Rapide to West Raynham to see the prototype Swift, Hunter and DH.110 but the Sabre was the most impressive of the lot in flight. Later, a farmer near Witney in Oxfordshire found the elevators from the Javelin – the RAE said that it was flutter at the highest rate ever recorded at 22 cycles per second.

A contributory factor in Waterton's landing accident was the fact that the nose-wheel leg on the Javelin was relatively long in comparison to the main wheel legs so that in the ground attitude the fuselage and wing were presented to the airflow at a positive angle of incidence [without rearward stick movement the Javelin would leave the ground at around 100 kts IAS]. As the touchdown also had to be made at a speed that was much higher than normal, in gusty conditions and with extremely limited pitch control, the landing was thus made exceedingly difficult. For his outstanding bravery in retrieving the auto-observer records from his burning aircraft Bill Waterton was later awarded the George Medal.

Despite the loss of WD804 the Ministry of Supply awarded a production contract for the GA.5 on 7 July 1952 with 'super-priority' status and it was at this time that it was given the name Javelin. The rival DH.110 dropped out of favour with the RAF as it was felt that the Javelin had greater development potential, although the de Havilland aircraft was to be transformed into the Sea Vixen for the Royal Navy. Test flying on the Javelin was resumed after the first flight of the second prototype (WD808) on 21 August 1952 and its participation at the SBAC Display at Farnborough in early September. It was joined by the third aircraft (WT827) on 7 March 1953 and this was the first Javelin to be fitted with the required armament of four 30 mm cannons mounted in the wings and AI radar in the nose. Thus far the Javelin had been flown with a rounded nose but it was found that the neoprene skin applied to the blunt Hycar radome quickly deteriorated in rain and eventually, after much trial and error, a more pointed nose radome was developed.

Despite its size and generous wing area (927 sq.ft) the Javelin had internal space for only 765 gallons of fuel which led to sortie times of around fifty minutes. Bill Waterton had already made the point that this was wholly inadequate and his view was backed up by Colonels Dick Johnson and Pete Everest who flew the Javelin in March 1953 as part of an evaluation carried out by USAF test pilots under the Mutual Defense Assistance Program. WT827 was flown with two 250-gallon ventral tanks in June 1953 however the weight and drag of these external tanks affected performance, especially above 40,000 ft [the shortage of internal fuel would not be addressed until the advent of the Javelin FAW.5 as related in Chapter Seven].

As the Javelin was likely to spend much of its time at high altitude its manoeuvrability in this region was paramount. Early testing had been a little disappointing in this respect but developments with models in the wind-tunnel had shown that instead of a leading edge with constant sweep, a reduction in wing sweep on the outer portion of the wing would improve manoeuvrability at high altitude. Sweepback was therefore reduced outboard of the cannon installation and this had the effect of reducing thickness/chord ratio on the outer section. The new wing shape was flown for the first time on WD808 on 28 May 1953 however this aircraft was lost in an accident on 11 June that killed test pilot Peter Lawrence.

The last two Javelin prototypes were WT830 which was flown for the first time on 14 January 1954 and WT836 which joined the test programme on 20 July 1954. By this time Bill Waterton had left Glosters and had been replaced as Chief Test Pilot by W/C R.F. 'Dicky' Martin DFC AFC who had previously been an instructor with the Empire Test Pilots' School at Farnborough. WT830 was the first Javelin to feature fully powered ailerons. During the development of the Javelin the canopy design had shown a number of revisions from that fitted to WD804 which had featured an enclosed rear cockpit with two small 'port holes'. It had been assumed that the radar operator/navigator would be able to work better in a dark environment but after a number of changes, including square windows in WT827, a fully glazed canopy was adopted for both pilot and navigator on WT836 [this aircraft was also used for hood jettisoning trials at A&AEE Boscombe Down]. Even the glazed canopy had to be modified as it had to be raised slightly so that there was sufficient room for the crew's bone domes and for them to be able to reach up for the ejection seat firing handle.

The first production Javelin FAW.1 was XA544 which was taken into the air for the first time on 22 July 1954, only two days after WT836. It was powered by two Armstrong Siddeley Sapphire Sa.6 engines of 8,000 lbs thrust at sea level and was retained by Glosters for trials work, both at their airfield at Moreton Valence and with RAE and A&AEE. In fact half of the production run of Javelin FAW.1's never saw squadron service with the RAF as they spent their entire lives either with the manufacturers or with various research establishments, which only serves to highlight the short-sighted policy of employing such a small number of proto-types on what was an extremely advanced and complicated aircraft programme.

CHAPTER 2

Testing the Javelin FAW.1 and FAW.2

The fourth prototype Javelin (WT830) underwent a brief series of pre-view handling trials by pilots of A&AEE Boscombe Down commencing on 28 June 1954. The aircraft was still subject to a number of flight limitations at the time, and no stalling or spinning was conducted. The determination of manoeuvre boundaries was also not attempted and very little transonic experience was gained, in particular aileron was not applied at higher Mach numbers. Take off weight with a full internal fuel load of 740 gallons was 29,300 lbs with a normal landing weight of 25,400 lbs.

On the ground, idling thrust was insufficient to get the aircraft moving once the wheel chocks had been removed so power had to be increased to be able to taxy. Once underway the undercarriage proved to be soft and spongy and gave a comfortable ride. Lining the Javelin up for take off was straightforward and with flaps up, tailplane set to 1 division nose up and full fuel, the engines were run up against the brakes to 7,500 rpm. There was a tendency for the nose to drop at this point but the brakes were effective in keeping the aircraft stationary. The brakes were then released and the throttles opened progressively to the fully open position to give 8,600 rpm. There was hardly any swing and acceleration was rapid. Directional control was maintained initially by slight use of differential brake but once 60 kts IAS had been reached the rudder became effective and this control was used thereafter to keep straight. The nose wheel lifted at about 90 kts IAS and the aircraft was airborne at 130 kts IAS. A slight backward pressure on the stick was required to lift the nose but the aircraft flew off in trim with the stick position approximately central.

Once airborne the brakes were applied and the under-carriage retracted, the retraction process taking around 7½ seconds from pressing the button. Although the aircraft tended to wallow slightly at lower airspeeds, control forces were light with adequate response and there was no appreci-able change of trim when leaving the ground or retracting the undercarriage. The take off distance was short at around 600 yards and acceleration was brisk during the take off roll and subsequent climb out. Take offs were made with varying tailplane trim settings from 1 division nose down to 3 divisions nose up and in each case the out of trim setting could be easily held on the controls up to 200 kts IAS, the control forces being 10–15 lbs.

For landing the circuit was joined at 300 kts IAS at a height of 2,000 ft and with the engines throttled to 6,000 rpm. When the airbrake was selected there was appreciable deceleration but there was no trim change and negligible buffet. With the airbrake out speed rapidly decreased to 165 kts IAS at which point the undercarriage was lowered, the wheels taking around ten seconds from pressing the button to the cockpit indication of three greens. Power was then set to 6,800 rpm which gave a comfortable circuit speed of 165 kts IAS and in this condition the tailplane was set to 2 divisions of nose up trim. At this speed the ailerons were light and responsive but there appeared to be a 'dead period' around the mid-point of the elevator which gave a spongy feel. Half flap was selected when flying crosswind with full flap on finals, in each case there was little apparent trim change although there was mild buffet after selecting full flap. The view during the circuit and final approach was described as excellent.

The final approach was made at 130 kts IAS and the only criticism of the controls at this stage was that the rudder was light and not very effective and there was a slight deterioration in elevator response. As the airfield boundary was crossed at 120 kts IAS the engines were throttled back and the aircraft touched down at around 110 kts IAS. The landing was straightforward, the aircraft flying onto the runway with a slight nose-up attitude requiring a light pull force on the stick of approximately half way back from the central position. The brakes were applied soon after touchdown and there was no tendency to swing during the roll out. The landing run took approximately 1,600 yards with only moderate braking in conditions of light wind. A simulated flapless approach was also attempted. The aircraft was trimmed on the approach with undercarriage down at 130 kts IAS with 5 degrees nose-up trim and 6,000 rpm. This resulted in a long flat approach during which it was found that it was easy to build up speed by opening the throttles but it was much more difficult to lose speed and so careful manipulation of the throttles was essential to maintain a reasonable airspeed. It was found that a slightly higher approach speed (5–10 knots) was necessary for optimum control in the flapless condition.

For an overshoot the aircraft was brought onto finals with wheels and flaps fully down, 150 kts IAS at 6,300 rpm and with the tailplane set to 2.5 divisions nose-up. Speed was gradually reduced to 125 kts IAS at 400 ft and 120 kts IAS and 6,800 rpm at 300 ft. When the aircraft had descended to 100 ft the throttles were opened to 8,000 rpm and the nose was raised for the overshoot. The increase in power brought about an almost immediate acceleration and speed built up rapidly. The undercarriage was then raised and the flaps moved to the half flap position. During this process the loss of height was minimal (approx 20–30ft) and acceleration was

14

rapid so that in spite of throttling back, speed had built up to 180 kts IAS before the flaps were fully raised. The aircraft was very easy to control throughout the overshoot and any out-of-trim forces were light and easily held on the controls. With the excellent level of acceleration shown it was felt that if higher power settings had been used on the overshoot the acceleration might then have been so rapid that the aircraft could easily have exceeded the airspeed limits for flight with undercarriage and flaps down.

A brief check was made of the Javelin's performance and handling at high indicated airspeeds. With the engines set at 8,000 rpm and the aircraft trimmed for level flight at 1,000 ft, there was little difficulty in reaching 500 kts IAS. Power was then increased to 8,300 rpm and speed gradually built up to 550 kts IAS which was the maximum limiting airspeed at that time. Although there were limitations on the amount of aileron angle that could be used it appeared as though aileron control was quite adequate with light stick forces for small displacements [aileron movement was limited by restrictors at speeds above 410 knots so that full stick movement did not produce full aileron movement]. The tailplane trim was extremely effective, so much so that the rate of operation was such that even a short blip was likely to impose a substantial jerk on the aircraft. The elevator control was particularly heavy however and although no exact measurements were taken it was estimated that stick force per 'g' at 550 kts IAS was in the order of 30 lbs per 'g'. The most impressive feature of the Javelin was the ease with which an indicated airspeed of 550 knots was achieved and even at 8,300 rpm (300 rpm below full power) the aircraft was still accelerating in level flight.

During the time when the aircraft was being flown at high speed a brief investigation was made into the effectiveness of the airbrakes. These were selected fully open in a shallow dive at 2,000 ft, 500 kts IAS and with the engines set at 7,800 rpm. Initially there was little deceleration and speed decreased only gradually. This situation continued until a speed of 430 kts IAS was reached when there was some improvement and a more significant increase in deceleration occurred at 410 kts IAS. It soon became apparent that this uneven deceleration was caused by the airbrakes not opening fully at the higher airspeed but at no stage was there any noticeable change of trim or buffet. Prior to the test by A&AEE the manufacturers had reported the phenomenon of 'duct banging' when the engines were throttled from high airspeeds. This was also experienced during the trial and was described as a series of popping bangs that were similar in effect to the exhaust popping of a Rolls-Royce Merlin engine when the engine was throttled for landing. The more the engines were throttled back the worse the banging became but the investigation was not continued in order to avoid damage.

Longitudinal stability was tested at 40,000 ft by trimming the aircraft in level flight at 0.85 IMN. A turn to port was initiated and this was gradually tightened. This produced a loading of 2g which was reached with ease and with light stick forces in the order of 5 lbs. The turn was tightened to 2.5g at 0.80 IMN when slight airframe buffet became apparent but there did not seem to be much increase in stick force, if any at all. The stick position in the turn was approximately two thirds back from the central position and as a result of this it was felt that stick travel might be the limiting factor at altitude.

Checks were also made on the Javelin's oscillatory stability during which it was noted that it had a tendency to fly right wing low (approx 3 degrees). From straight and level flight at 10,000 ft and 200 kts IAS, a 5 degree yaw to port was initiated, requiring a 10–15 lbs foot force on the rudder. The controls were then released and there followed a moderately well damped directional oscillation which was damped out in four cycles. The same test was tried to starboard with similar results except that the cycles were larger in amplitude and required five oscillations to damp out. At higher airspeeds (400–420 kts IAS) it proved to be impossible to set up a

5 degree yaw owing to the heavy rudder loads. Maximum foot force only gave some 2 degrees of yaw and directional oscillations were quickly damped out.

Although the Javelin had shown itself to have excellent stability characteristics at lower levels, this was not the case at higher altitudes. At 45,000 ft the aircraft was flown at 200 kts IAS with the engines set at 7,500 rpm. A 5 degree yaw to starboard was set up before the controls were released but on this occasion the motion was undamped after five oscillations, by which time the nose was still moving 2–3 degrees either side of centre. There was no tendency for a wing to drop although some lateral motion was noted which was in sympathy with the directional oscillation. This lateral accompaniment had not been seen at lower levels. After five oscillations the nose had dropped by 5 degrees and the wings were dropping alternately 2–3 degrees. When the same was attempted to port the directional oscillation was accompanied by the starboard wing dropping rapidly and at the same time the nose went well down. By the third oscillation the starboard wing had dropped 90 degrees and with the nose also dropping, normal control was resumed.

The rate of roll was checked at varying airspeeds at 10,000 ft during rolls to port using maximum aileron deflection and also a certain amount of rudder. The times taken to roll through 360 degrees varied from 3.8 seconds at 200 kts IAS to 2.5 seconds at 400 kts IAS. At the latter airspeed a harsh airframe buffet was felt at maximum aileron deflection which appeared to come from the wing tips and persisted until the aileron control input was reduced. This phenomenon was later discussed with technical staff from Glosters who put the problem down to aileron stall.

An attempt was made to discover the amount of time the Javelin took to reach high altitude in a full power climb, although as a turn had to be made through 180 degrees after take off, this was not fully representative of true performance. Even so, it proved to be a sprightly performer passing 10,000 ft in 3 minutes 5 seconds and 20,000 ft in

4 minutes 25 seconds. The Javelin went through 30,000 ft after 6 minutes 15 seconds and eventually reached 45,000 ft in 11 minutes 5 seconds, figures that were significantly better than the Hunter F.1. At 45,000 ft the Javelin was trimmed into level flight with 1 division of nose-up trim at full throttle. With the port engine giving 8,400 rpm and the starboard 8,300 rpm, speed gradually built up and eventually stabilised at 0.875 IMN which, when corrected, gave a true Mach number of 0.91. Aileron control was good and the aircraft was stable laterally but directional damping was poor. The view from the cockpit at altitude was also disappointing, especially when looking down. This was because of the wide cockpit coaming.

The Javelin's behaviour at high Mach numbers could only be assessed up to the limiting factor at the time of test (0.93 IMN), but in general the aircraft proved to be remarkably docile. The effects of compressibility became apparent at about 0.87 IMN with the gradual onset of a nose-down change of trim. An increasing pull force of 10–15 lbs was needed to maintain level flight and slight airframe buffet was felt at 0.90 IMN. As speed was built up to 0.92 IMN a slight lateral instability was noted with mild wing rocking. The limiting speed of 0.93 IMN was reached with no difficulty and with only a slight nose-down attitude it became clear that higher Mach numbers could be reached without undue effort. The recovery from this speed was straightforward, with the aircraft again running through the lateral instability and longitudinal trim change. A simulated letdown was made from 45,000 ft, initially with full airbrake selected but this gave an excessive nose-down attitude which could not be followed on the artificial horizon. As a result half airbrake was set, and with an engine setting of 7,000 rpm and 250 kts IAS, this gave a very stable instrument platform for descent together with an adequate rate which averaged approximately 7,000 ft/min.

Flight at slow airspeed was checked at 10,000 ft with the wheels and flaps up, tailplane trim set to 4 divisions nose-up, all control forces trimmed out and the engines set to 6,000 rpm. At 130 kts IAS a slight airframe buffet was noticed which gradually increased as speed was reduced. As airspeed was further reduced an increasing pull force was needed to hold the nose on the horizon until at 115 kts IAS there was some easing in the amount of back pressure required. At this speed the rudder was light but its effectiveness was poor. In contrast the ailerons were still effective but the response was not as good as it had been at higher airspeeds. Speed was brought back to 110 kts IAS and to keep the nose on the horizon the stick had to be slightly back from central with a light pull force of 3–5 lbs. There was no apparent inclination for the nose to rise, nor was there any evidence of stick-force reversal. On completion of the test the throttles were opened and the aircraft accelerated away.

Despite the limited scope of the trial due to the flight restrictions that were still in place, the Javelin made a favourable impression especially with its excellent level of performance and docile handling characteristics. The main recommendations made by A&AEE were for a reduction of the high elevator control forces and the elimination of the sharp yaw and buffet that was experienced at transonic Mach numbers. Another aspect of the Javelin's handling that needed improvement was its lateral and directional stability at low airspeed and high altitude. It was also felt that the internal fuel capacity was insufficient for an all-weather fighter.

The Javelin FAW.1 was also tested by the Central Fighter Establishment at West Raynham in a trial that commenced on 10 January 1956. The aircraft used were XA565, XA566 and XA568 and a total of 258 hours were flown of which 34 hours were at night. In general the Javelin made a favourable impression although there were several aspects of its operation that left a lot to be desired. One such was engine starting, as the Mark 12 starter cartridge proved to be extremely unreliable and during the course of the trial there was a 30 per cent failure rate. There was also a wide divergence with regard to scramble times depending on the aircraft's readiness state. With the crew already in the cockpit, scramble times from the ORP averaged just under a minute; at night however when scrambling from a lower state of readiness the lengthy strapping in time, and difficulties manhandling the cockpit access ladder, meant that scramble times were more like six minutes.

During the trial CFE pilots commented on the fact that the Javelin FAW.1's elevator control was rather insensitive below 250 kts IAS which, combined with 'impressively effective airbrakes', meant that careful attention had to be paid to airspeed when flying in the circuit or during an instrument procedure. This was not considered to be a good feature for an all-weather fighter and it was recommended that the elevator feel at low speeds should be improved. In terms of manoeuvrability the worst aspect of the FAW.1 was the early onset of buffet in turns and a stall warning device was an urgent requirement. Acceleration became progressively worse with increase in altitude so that at 48,000 ft it was impossible to accelerate from 0.70 IMN in level flight. This made it necessary to maintain high speed (at least 0.85 IMN) until late in an interception because if speed was allowed to fall away, acceleration to a reasonable overtake speed took a long time. CFE pilots were also scathing as regards the FAW.1's internal fuel capacity which was described as 'quite unacceptable' with average sortie times of only 45–50 minutes.

Although next in number sequence, the Javelin FAW.2 actually entered RAF service a few months after the FAW.4 and FAW.5. It was tested at CFE in the summer of 1957 with XA808 and XA809 being delivered to West Raynham on 21–22 May, however a third aircraft (XA806) was delivered on 29 July as a result of slow progress due to persistent unserviceability. This included five instances of engine flame out which led to the first two aircraft being returned to the manufacturer for investigation of the problem [it was eventually discovered that the flame outs were caused by defective air shut-off valves in the No. 3 fuel tanks]. The biggest difference between the FAW.1 and FAW.2 was the latter's American-designed AN/APQ-43 radar (British designation AI.22). This also caused severe problems and a tactical evaluation was not possible due to difficulties with the track set of the radar. Use of AI.22 resulted in an increase in all-up weight but the fuselage forward of the pilot's cockpit was fifteen inches shorter. The nose radome on the FAW.2 was also hinged to allow access to the radar installation [on AI.17-equipped Javelins the radome had to be removed to allow servicing to take place].

The climb performance of the Javelin FAW.2 was evaluated in climbs up to 48,000 ft and it was discovered that it was inferior not only to the FAW.1 and FAW.4, but also to the FAW.5 which had increased internal fuel capacity and weighed 1,700 lbs more on take off. The average time for the FAW.2 to reach an altitude of 40,000 ft was 10 minutes 22 seconds which was 2 minutes 30 seconds longer than the FAW.1. No obvious reason could be found to explain this discrepancy, although it was thought that the aircraft's altered nose profile might have affected its aerodynamic qualities. Acceleration at 45,000 ft was also worse and the FAW.2 took 1 minute 48 seconds to accelerate from 0.85 IMN to 0.90 IMN which was twice as long as the FAW.1. Maximum speed was also slightly lower at altitude (0.89 IMN as opposed to 0.90 IMN) but turn performance was similar.

Total production of the Javelin FAW.2 was only thirty, of which twenty-six were delivered to the RAF where they served with 46, 85 and 89 Squadrons. The service life of the FAW.2 was extremely short with 85 and 89 Squadron receiving FAW.6's by 1960. The last Javelin FAW.2's were flown by 46 Squadron which disbanded in June 1961

CHAPTER 3

Into Service

The Javelin entered RAF squadron service on 24 February 1956 when FAW.1 XA570 was delivered to 46 Squadron at Odiham from 23 MU at Aldergrove. The arrival of the new all-weather fighter created much anticipation and the station commander at Odiham, G/C A.K. Gatward DSO DFC, was on hand to accept the new aircraft. It was delivered by a ferry crew comprising F/Sgt Bob Hillard and F/Sgt Gray Murrin but the assembled masses were unaware that the Javelin had arrived on time thanks to a small ladies wrist watch. Bob Hillard's RAF watch had broken and he had been forced to borrow his wife's watch which had no second hand and was so tiny that it almost needed a magnifying glass to read. As Murrin had forgotten his watch this small timepiece had to be used to ensure that the timekeeping en route was correct and the Javelin arrived at Odiham on schedule [Bob Hillard's logbook shows that he flew three different types of aircraft for the month of February 1956, the Auster 6, Anson 19 and Javelin FAW.1!].

Over the next few months new Javelins arrived at the rate of 4–5 per month so that by May the unit's strength was up to fifteen. At first the serviceability rate was 'appallingly low' and pilot conversion slowed to a crawl but there was a gradual improvement so that by the end of May 170 hours had been flown during the month by day and 56 hours by night. After the pilots had been converted to the new aircraft, navigators took centre stage as they came to grips with the AI.17 radar. Exercises in airborne interception were carried out with the aid of GCI stations and night flying was increased to four nights per week. This rapidly improving situation came to an abrupt halt on the night of 11/12 June 1956 when 46 Squadron's CO, S/L F.E.W. Birchfield and his navigator F/O B. Chambers, were killed when their Javelin (XA570) hit the ground during a ground controlled approach.

They had taken off at 0022 hrs using the call-sign 'Warmtoast 70' and had carried out a practice interception at 35,000 ft with another Javelin flown by S/L G.I. Chapman, the CO of the Javelin Mobile Training Unit (JMTU) which had temporarily moved to Odiham to assist with the conversion process. After completing the training exercise S/L Birchfield requested a QGH/GCA procedure. Weather conditions at the time were 5/8 cloud at 1,100 ft agl, a visibility of 2,200 yards with a wind direction of 340 degrees at 9 knots. There was also light intermittent rain. During a normal instrument approach the aircraft would have been picked up at around 12–15 miles by which time its height should have been 3–4,000 ft. On this occasion however the first radar contact was at six miles with the aircraft at 2,000 ft. This was too close for the GCA controller to accept and so Approach control passed QGH information. At five miles range the Javelin was cleared down to 1,000 ft which was acknowledged. At 1¾ miles range Birchfield advised that he was in visual contact with the airfield but when asked if he would like a low level GCA he replied '70. I'm ...' at which point the transmission ended abruptly.

When the remains of the Javelin were found it was discovered that the wreckage trail extended for only 400 yards and there was practically no 'dig in' suggesting that the aircraft had hit the ground with an excessive rate of descent.

During trials at the Central Fighter Establishment it had been noted that at approach speeds the elevator control was rather insensitive which, coupled with very effective air-brakes, meant that it was necessary to pay careful attention to speed control, especially during instrument approaches. The normal practice was to approach with flaps up and air-brakes out at 6,700 rpm but with this configuration and power setting a pronounced sink was possible almost approaching a state of mush and it was thought that this may have occurred on this occasion. The aircraft had hit the ground with the undercarriage still retracted.

No. 46 Squadron's new CO was W/C Harry White DFC AFC who was a Second World War night-fighter 'ace' having shot down twelve German aircraft. June was a difficult month from a maintenance point of view as extensive modifications were carried out including work to hydraulic pipelines, the fixing of stronger rivets in the air intakes, double engine changes and alterations to the electrical system in some air-craft. Apart from the engine changes the work was mainly undertaken by the appropriate firms and involved a large number of man hours. In addition the Javelins were grounded twice for short periods due to irregularities in the hydraulic systems and investigation by the manufacturers (Dowty). The aircraft situation was not helped when F/O Tony Warner had the undercarriage collapse on XA624 after landing, resulting in Cat.3 damage. As a result of all these difficulties the number of aircraft serviceable on any single day was no more than four.

Serviceability did improve towards the end of the year and it was just as well as the squadron flew Intensive Flying Trials commencing in August which involved flying 1,000 hours. The aim was to fly 480 hours per month using eight trials aircraft and this was completed successfully on 4 October. The squadron took part in Exercise Beaverbrook on the 19th and claimed six B-36D bombers which were flying at 24–40,000 ft. The following month some of the unit's Javelins took part in a trial with a Canberra from RAE

Farnborough which was dropping window (chaff) to see how it affected the AI.17 radar. Rather perversely the Javelin that was directly astern of the Canberra managed to hold lock better than those flying at either side.

In January 1957 the first night exercise in which Javelins operated against V-bombers took place when aircraft of 46 Squadron deployed to Church Fenton and West Malling. Valiants were intercepted at altitudes of 40,000 ft + with excellent results. In February interceptions of Valiants were made at 48,000 ft but operations at such altitudes caused problems as the airbrakes on some aircraft would not operate due to icing (this snag usually rectified itself by the time the aircraft had descended to lower level). On 18 March F/L P.A. Gifkins was carrying out cine gun attacks in XA624 at 24,000 ft when he inadvertently spun off a steep turn. The Javelin had already acquired a bad reputation for its spinning characteristics (see Chapter Eight) but by taking precise recovery action in accordance with Pilot's Notes he was able to regain control at 15,000 ft.

One of the highlights of 1957 was 46 Squadron's participation in a live BBC broadcast called 'This Is Your Air Force Now' in May. Rehearsals had taken place the previous month and involved the Javelins intercepting Valiants in mock attacks which were filmed from the back of a Beverley transport aircraft. Despite the large speed differences between the participating aircraft and the camera aircraft the interceptions were successfully filmed and broadcast. On 21 August what was to be a recurring theme with the Javelin took place when the port engine of XA618 was being started. The engine starter disintegrated and the debris punctured the No. 1 fuselage tank. Escaping fuel was ignited by the hot gases and sprayed onto the ground beneath the port wing, the fire extending along the length of the aircraft. The port side of the Javelin was badly burned before the fire section was able to douse the flames (XA618 survived the experience and later flew with 87 Squadron).

On 30 July 46 Squadron received its first Javelin FAW.2 and more arrived over the next couple of months to replace the FAW.1's. It featured the 'all-flying' tailplane and also had AI.22 radar which on the rare occasions when it did work was invariably unserviceable by the time the Javelin had returned to base. Once again there was a worrying shortage of spares and test equipment and on at least two occasions it was discovered that the radar installations had been wired up wrongly when the aircraft had been at Maintenance Unit prior to delivery. Around this time a number of new crews joined the squadron and one of the new boys was navigator Flying Officer Colin Edwards

I was trained on the AI.21 radar [Meteor NF.12/14] and joined 46 Squadron as they were converting onto the Javelin FAW.2. We grew to like the Javelin, but I never heard of a pilot who loved it like they did the Hunter or the Sabre. My pilot at the time was very nervous of the aircraft as there had been several recent fatalities. He and other timid pilots were particularly scared of stalling and so used to do long approaches more suited to Heathrow than Odiham. It was an amazing contrast to fly with the better pilots. I flew with Harry White soon after I joined the squadron and the difference was remarkable. His circuit seemed to be a continuous turn with the aircraft levelling its wings at the last moment. He flew much slower and like many of the experienced pilots would switch the 'stall warners' off. These were new on the FAW.2 and played a warning tune when the speed was low or when a high speed stall was imminent. I had a lot of fun scaring my pilot to death by playing an identical tune on a mouth organ. Looking back I suppose this did little for his confidence.

We had a lot of trouble with AI.22 which, if I recall, was due to failure of the 'lock-on' system which never worked well. The maximum range of the radar was around twenty miles at altitude but down to half that or

26

less at low level. The AI.22 system had a PPI indication which was helpful if we ever wanted to map read but this was only really possible for coast lines and did not make much difference to our interception work where we were concerned with angle off and range. The AI.22 did have the ability to restrict its 'beam' to a very narrow width and height giving a much faster blip repetition rate. In theory this was very useful in following an evading target at close range. The most extreme example was 'oriental sector scan' when the beam could be cut to five degrees which illuminated the target in a sharp torch-like beam. This was fine when you were straight and level but as soon as you banked it was necessary to move the scanner (this was supposed to be overcome by the lock-on of AI.22). For most this was impossible.

There was a legendary American instructor at OCU who could do it. Instead of the patter which we were taught to use as commentary for our pilots, he had a different procedure. He checked with each pilot as to where they wanted their visual – most wanting about 12 o'clock at five degrees up. The pilot then received little or no indications, merely orders and the constantly reassuring 'Looking pretty good ... looking pretty good.' The pilot would be sweating away weaving, diving and climbing in the dark not knowing how far away the target was until this 'ace' American said 'Look up buddy, there's your target.' I believe he had used AI.21 a lot in the States.

Our visuals might be anywhere from 800 yards down to 200 yards. I seem to recall Valiants as being particularly difficult to see and remember the luxury of being able to take my head out of the visor to look up at the Vulcan we had just intercepted. Even though we were right underneath it I had to work out its shape from where the stars weren't. On exercise we must have flown very close to targets particularly on head-on interceptions (which were normal) because we had to do a blind turn

27

from eleven miles and often felt the bump as we went through the slipstream of the target.

How effective we would have been in anger is difficult to say. I can remember our boss, Harry White a triple DFC night-fighter ace, saying that few of us would get in behind our target and even fewer would shoot it down. We could certainly have intercepted all the Russian bombers in service at the time, providing they played the game and came above 10,000 ft and didn't evade too much. We believed that they would have to come fairly straight for the last 100 miles where we would be getting them. We had very powerful guns and I feel the boss was remembering the shortcomings of the Second World War night-fighters and their ground control. We practised operating without ground control frequently; this would be either 'loiter' or 'broadcast' systems. We also practised against jamming targets, usually American B-66s.

We had few problems with the first generation of V-bombers but had to be very careful with the Vulcan and Victor B.2s both of which would be able to run away from us. It was essential that we rolled out very close behind if we were to use our guns. This was not so much of a problem once we got Firestreak. Canberras could embarrass us by flying too slowly. They would weave at high altitude making it impossible for us to hang in the air within gun range. Low level interceptions by night would have been very difficult as the ground returns obscured the target. Our orders were to ram the enemy if we had used all our ammunition – so much for our contempt of Japanese suicide pilots. I also remember an intriguing order to surround the target!

One of the chores of flying in the days before the twin-stick Javelin T.3 was flying with new pilots on their first solos. If that was not bad enough, the thought of a visiting Air Vice-Marshal or station commander wanting to fly could clear the crew room of navigators in seconds. They found taxiing particularly difficult as they were unused to wheel brake steering. As an aside it was possible to 'fly' a Javelin from the back seat. If we undid the panel on the right of the cockpit we exposed the control rods and this could be used to give them a playful tug when the pilot was on approach. This was always useful if he was wasting valuable drinking time by insisting on practising circuits. We could also do barrel rolls. These were made easier by jamming the crowbar in the controls, thus having our own joystick. It was noticeable that the best pilots were very often ex day fighter pilots. They were more eager to see what the aircraft could do. Although loops were banned my next pilot, an ex Hunter and Sabre pilot, soon decided to see what all the fuss was about and on a dark night we first did a roll off the top and then full loops – all with no problems. As a day fighter we struggled to beat Hunters or Sabres but we could out-dive them using our massive airbrakes to slow us down.

I can honestly say that I was never frightened on the Javelin, although I was very often excited. During my time with 46 Squadron we had what was called 'Lightstrike'. As I understand it this was part of service politics when the RAF tried to prove that we could intercept fast moving MTB or E-boat type vessels, thus forestalling moves to equip the Navy with more aircraft. The game was that the boats would be found by maritime aircraft such as Shackletons or Dutch Neptunes. We would have been lurking at low level on one engine and would be called in at 2,000 ft to find the target. We usually picked up the boat at about six miles and, as I remember, the pilot started a dive at one mile with the target illuminated by flares dropped by the Neptune behind the boat. As soon as the pilot was visual we would transfer our attention to the altimeter shouting out the heights as they rapidly decreased. I think we were supposed to level out at about 400 ft. I do remember during some daytime practices continuing to call the heights out right down to the deck whilst the pilot, unbeknown to me, pressed home the attack. I would have continued to call the depths in fathoms!

One of the most dramatic moments during my time on 46 Squadron was when I was sat next to another Javelin at Sylt during an air firing camp. The usual happened on start-up and flames immediately shot out from just behind the navigator's cockpit. We waved all sorts of signals at the crew and they waved happily back not realising they were on fire. Most of the aircraft on the line were armed and so there was some panic whilst the neighbouring aircraft were moved. In the end we watched the aircraft burn out with great feelings of sorrow because it was loaded with several bottles of whiskey for the trip back to the UK. All was not lost however. When everything had cooled down we found a bottle of dimple Haig on the tarmac. It had fallen several feet without breaking [this incident involved Javelin FAW.2 XA802 and occurred on 9 March 1959]. I remember returning from that detachment unable to use my radar because the visor frame was a perfect wine rack for three bottles. It was a hot day and we did our usual 400 knot 4g break at Leuchars where we had to land for customs clearance. I was airsick in the circuit and threw my sick bag out of the aircraft as we taxied in. This aroused the customs officer's suspicions and I was delighted to take him to retrieve my contraband. Smuggling was very easy with an aircraft with lots of panels marked 'Weapons' etc, and our ground crews knew that we might hand them a watch together with our ejector seat pin.

I have very vivid and happy memories of my time on Javelins. There were wonderful drunken detachments but seemingly endless hours on standby. We used to provide year-round standby in the cockpit, ready to go. This was usually at Waterbeach or Wattisham and it was often freezing cold. To try to keep warm I used to take my holdall full of clothing so that I could put my feet and legs in it to stay warm. One day our telebrief to the GCI station became crossed with the local butcher's telephone and we spent the day receiving orders for sausages etc! We really were the forgotten few.

CHAPTER 4

The Javelin FAW.1 Described

Although for the 1950s the Gloster Javelin had a futuristic shape its construction was entirely conventional and it was almost exclusively made of aluminium alloy with a small amount of steel. The following information relates to the Javelin FAW.1 and FAW.4.

The fuselage was built in four sub-assemblies, the two largest (the front and centre fuselage) being permanently joined together during manufacture. The fuselage nose and the rear fuselage were easily detachable, the latter to give access for engine removal. The fuselage nose was constructed of fibreglass reinforced plastic as it enclosed the AI scanner. The pilot and navigator were accommodated in a pressure cabin which was formed in the front fuselage by an inner shell which was sealed at the front and rear by solid frames and above by the windscreen and electrically-operated sliding hoods. The engine air intakes were built onto the sides of the front fuselage with ducting connecting them to the engine compressor casings.

The joint between the front and centre fuselage sections was at frame 12 which was a large fabricated light-alloy structure that also carried the attachment points for the main plane spars. Running fore-and-aft through the centre fuselage was a deep box-beam structure made up of the centre portions of the frames and two longitudinal diaphragms. The outer portions of the frames were attached to each side of this structure to form the two engine bays. A servicing bay was located centrally in the underside of the fuselage to provide space for the starter equipment and services such as a hydraulic panel, generator control and electrical distribution. The rear fuselage was of semi-monocoque construction which incorporated the jet pipe exits. Two pressure heads were located on the upper rear fuselage to serve the tail plane feel simulators.

The delta configuration main planes were each constructed about a heavy main spar and a lighter rear spar, the inboard end of each forming the attachment to the fuselage. The main structural element of the wing was a torsion box formed by the main spar, the leading edge ribs and the light-alloy skin which ran from root to tip. Flaps and, further aft, the airbrakes were embodied towards the wing roots with lateral control being provided by internally mass-balanced ailerons. The wings of the FAW.4 also featured vortex generators on the upper surface to re-energise the boundary layer airflow [in squadron service these proved to be a particular hazard for the nether regions of any armourer unfortunate enough to lose his balance and slide backwards off the wing!]. A pressure head was fitted on the port outer main plane which provided pressure and static supplies for the airspeed indicators in both cockpits and the Mach meter in the pilot's cockpit. It also provided static supplies only for the altimeters, the rate of climb and descent meter, the altitude unit of the GGS control unit and the zero-reader flight computer. The tail plane was mounted on top of the swept fin which was attached to the centre fuselage. The rudder and elevators were all-metal, internally mass-balanced structures, the rudder having a combined balance and trim tab.

The undercarriage was hydraulically operated and consisted of two independent main-wheel units which retracted inwards into recesses in the main plane and a castering nose-wheel unit that retracted rearwards into a recess under the pressure cabin. All three units had liquid-spring shock absorbers which were automatically recuperated from the hydraulic system. The port undercarriage wheel bay was the location for a tele-briefing landline plug for use when the aircraft was at cockpit readiness. Braking was by hydraulically-operated disc brakes fitted to the main-wheels.

The main flying controls were operated by a normal control column and rudder bar in the cockpit. The ailerons were power-operated by a hydraulic system, while the rudder and, in the FAW.1, the elevators were power assisted. Tail plane incidence was variable in flight and for the FAW.1 was electrically operated with normal and emergency switches being mounted in the cockpit. The tail plane of the FAW.4 was operated hydraulically by movement of the control column, the elevators serving as anti-balance tabs. A hand operated trim tab was fitted to the rudder.

The early Javelins were fitted with one Sapphire Mk.102 and one Mk.103 axial-flow gas-turbine engines fitted in the port and starboard engine nacelles respectively. Each engine developed approximately 8,150 lb static thrust at sea level. Rails were fitted in the centre fuselage so that the engines could be installed from the rear and their mounting trunnions were locked to the centre structure and to heavy mounting frames with a steady mounting provided under the compressor casing. A shaft from each engine drove an auxiliary gearbox mounted in frame 12. The engines were fitted with turbo starters, the cartridge holders also being mounted on frame 12. Fuel was carried internally in a number of bag-type tanks with five in each main plane and four in the centre fuselage. The fuel that was held in the wing tanks and the two forward fuselage tanks was transferred by air pressure into two collector tanks in the centre structure. From the collector tanks it was supplied to the engines by electric booster pumps which incorporated inverted flight valves. Total internal fuel was 6,120 lbs of AVTUR or 5,890 lbs of AVTAG [the difference in weights between the two fuels was due to their different specific gravity].

The power required to operate the hydraulic services was provided by three pumps driven by the gearbox. One pump operated the undercarriage, flaps, airbrakes, wheel brakes and leg recuperators. The other two pumps operated the flying control assistors with one also acting as a stand-by for the other services. In an emergency situation the under-carriage could be lowered by a compressed air system.

Power for the electrical system came from two 6,000-watt, 24-volt generators which, in addition to instrument and heating and lighting services also operated the tail plane actuator on the FAW.1 and the hood winding gear. Inverters supplied a.c. supplies for the radar, gun firing and flight instruments. The AI scanner was mounted on the foremost front fuselage bulkhead while its associated equipment was fitted in the AI bay, in front of the cabin. Other radar equipment included Gee, ILS and IFF. Communications was provided by two VHF radios fitted under the rear cabin floor. Two whip-type aerials, one for each set, were fitted to the top of the fin.

The four 30 mm Aden guns were mounted in gun bays outward of each main wheel bay. Each gun was mounted between two ribs and was belt fed from an ammunition tank built into the wing structure with 200 rounds per gun. A gyro gunsight and a collimator were mounted in the pilot's cockpit while a cine camera was installed in the starboard wing root nosing. Other equipment included combined hood and seat ejection, cockpit air conditioning, oxygen, wind-screen de-icing and an anti-g installation. The overall length of the Javelin FAW1 and FAW.4 was 56 ft 4 in and the wing-span was 52 ft. Compared to day fighters of the time the Javelin sat quite high on the ground and the pilot's eyeline was approximately 11 ft above ground with the delta tail plane 16 ft above ground. Sweepback on the inner plane was

39.43 degrees which was reduced to 33.8 degrees on the outer section. Total wing area was 928 sq.ft.

The main flight instruments were the HGU (attitude indicator), the Mk.4B compass, the turn and slip indicator and the zero-reader indicator, all of which were located on the centre instrument panel. These items were all electrically operated. Other instruments positioned centrally were the ILS and the radio altimeter. The latter had a single red low-level light in the front cockpit and red, green and amber limit lights in the rear cockpit. The red light normally came on when the aircraft was flown below the pre-selected height but also came on in the event of failure of the equipment. The transmitter and receiver aerials for the radio altimeter were located on the underside of the port engine nacelle, and the port wing adjacent to the root rib, respectively.

The engine starting controls comprised a booster-pump, circuit-breaker, master starting switch, ignition isolating switch and an engine starter button (one of each for each engine) and these were grouped together on the starboard console. Two throttle levers were mounted in a box on the port side of the cockpit. This was also the location for the HP fuel cock levers, the handles of which incorporated the relight buttons. The engine instruments comprised a dual JPT gauge, oil pressure gauges and per cent rpm indicators. Controllers for the two 10-channel VHF radios, together with a change-over switch were situated on a panel under the port cockpit coaming. When the tele-brief landline was plugged in the VHF circuit was disabled.

The Javelin was fitted with a fire detection system incorporating (in each engine) six re-setting flame detector switches around each compressor casing with a further six in the exhaust cone shroud. Two fire extinguisher bottles were stowed between the engines on the rear of the front fire wall. Should a fire cause any of the detector switches to operate, the appropriate engine fire warning light incorporated in the fire extinguisher push button in the cockpit would illuminate. When the push button was pressed the extinguisher serving that engine would discharge its contents through spray rings. If this operation was successful the warning light in the cockpit would then go out. There were also two inertia switches situated in the centre fuselage that operated the extinguisher bottles automatically in the event of a crash landing. In such a situation these also isolated the aircraft battery and generators.

The early mark Javelins were fitted with Martin-Baker Mk.3J ejection seats which incorporated two firing handles, a primary firing handle above the headrest, and a secondary firing handle on the front of the seat-pan. The seat was fitted with a type ZF harness, the Mk.9 parachute, a personal survival pack, emergency oxygen system and leg-restraining cords. The height of the seat could be adjusted by a lever on the starboard side of the pan. When either firing handle was operated the hood was jettisoned immediately and a timing unit on the back of the seat was started. This unit withdrew the sear from the seat gun one second later and ejected the seat. The drogue fired half a second after ejection. Should an aircraft be subject to 'g' at the time of ejection, thus forcing the crew to use the secondary firing handle, it was important that the occupant of the seat should press his head back on the headrest to prevent injury to the spine. All leads incorporated quick-release mechanisms which were auto-matically broken when an ejection was initiated.

As the seat was ejected a static rod set the barostatic time-delay mechanism. For ejections made above 10,000 ft a free descent would be made to this altitude whereupon the barostat removed an obstruction to the gear train of the mechanism allowing it to operate. For all ejections below 10,000 ft this operation was set in motion directly after ejecting. After a delay of 1¼ seconds automatic withdrawal of the parachute took place. As this was taking place the occupant was momentarily prevented from leaving the seat by two restraining straps until deployment of the parachute lifted him clear. A manual override D-ring was fitted over the ripcord D-ring and could be used to detach the parachute from the seat if it was necessary to operate the parachute manually. If this was attempted it was important that the D-ring should be pulled before releasing the safety harness as there was a danger of the occupant falling out of the seat with the parachute still attached to the seat.

CHAPTER 5

Tactical trials with the Javelin FAW.4

In typical British fashion, the Gloster Javelin appeared in a number of different marks but production in each case was relatively low. The FAW.4 was first flown on 19 September 1955 but the production run extended to only fifty aircraft and even this was split between Gloster and Armstrong Whitworth. Before it entered service a tactical trial of the Javelin FAW.4 was carried out by the All-Weather Development Squadron of the Central Fighter Establishment and four aircraft (XA735, XA749, XA763 and XA764) were collected from Moreton Valence with the first arriving on 15 October 1956.

The FAW.4 had the fully-powered 'flying' tail but in most other respects it was similar to the FAW.1, although it did have two rows of vortex generators on the leading edge of the outer wing panels, stall warning vanes and two pitot heads on the fuselage just forward of the fin. The all-up weight of a clean aircraft was 33,150 lbs which was about 500 lbs heavier than a Javelin FAW.1. Engine starting on the FAW.4 was by Mark 14 starter cartridge which proved to be slightly less troublesome than the Mark 12 of the FAW.1.

However the chance of a successful start at the first attempt was still not good due to faulty contacts and occasional bad cartridges.

Although the aircraft could be held against the brakes at full power for take off, this was considered to be unnecessary as take off distance without ventral tanks, with no wind and flaps up was well under 1,000 yards. With full ventrals the take off distance was slightly longer but was still less than 1,000 yards. During the initial part of the take off run a fairly heavy pull was required to centralise the control column but when the 'q' feel began to take effect at about 70 kts IAS, the stick centralised itself. With 0 degrees of trim the aircraft flew itself off the runway at 125 kts IAS (clean) and 128 kts IAS (ventrals). The time from releasing the brakes to take off was approximately fifteen seconds. There was virtually no trim change when the undercarriage was raised, however progressive forward trim was required as the aircraft accelerated to climbing speed. Acceleration was reasonably rapid with 400 kts IAS being reached just 1 minute and 12 seconds after brake release. Take offs were made with crosswind components of up to 25 knots without difficulty.

The best climbing speed was 400 kts IAS which was maintained until 0.85 IMN was reached at approximately 20,000 ft. In the clean condition the Javelin FAW.4 reached 20,000 ft in 4 minutes 9 seconds (4 minutes 41 seconds with ventral tanks) and was capable of reaching 40,000 ft in 9 minutes 22 seconds (11 minutes 10 seconds with ventrals). Although turns made below 20,000 ft made hardly any difference to these times, any turns made above 30,000 ft could significantly affect the time to height. When a comparison was made with the results of the CFE trial on the FAW.1 it was found that the climb performance of the FAW.4 was inferior to the earlier aircraft as it took 12 minutes 33 seconds to reach 45,000 ft compared with just 11 minutes for the FAW.1. This inferiority was even more marked above 45,000 ft where the climb rate of the FAW.4 fell away substantially.

The flying controls of the FAW.4 were fairly light and well balanced however they were somewhat heavier than the FAW.1 so that pilot fatigue was noticeably greater after gunnery exercises and high 'g' manoeuvring. The controls became progressively heavier with increasing speed and above 500 kts IAS restricted aileron movement meant that the rate of roll was considerably reduced. Elevator control, though heavy, was considered to be adequate, however the rudder was virtually immoveable at high speeds. There was further criticism of the FAW.4 in terms of tailplane trimming. The trimmer (which was operated by a switch on the control column) as well as being slower in operation than the FAW.1 was also more insensitive at all heights and speeds. The aircraft was difficult to trim accurately, especially above 40,000 ft, so that successful interceptions required a degree of concentration that soon became tiring.

The fitting of a stall warning device to the FAW.4 was warmly welcomed as it allowed the pilot much greater freedom of manoeuvre. Turns could be tightened well beyond the onset of buffet and some pilots expressed surprise at the degree of penetration into the buffet region before the stall warning was heard. Care had to be taken at high indicated airspeeds however as structural limitations could easily be exceeded before the audio warning operated. During the trial the possibility of pitch-up under high 'g' at speeds above 0.90M and in the height band 25–35,000 ft was thoroughly investigated. During simulated interceptions, turns were made considerably tighter than were likely to be necessary in service but at no time was pitch-up experienced.

The heavier elevator and apparently heavier ailerons made the FAW.4 more tiring to fly than the FAW.1 except at speeds above 0.92 IMN where the 'all-flying' tail was a definite improvement. The least satisfactory aspect of the Javelin's manoeuvrability was the rapidity with which speed dropped in turns or when the nose was well above the horizon. This was particularly noticeable when flying above 40,000 ft or when flying in the circuit at low power.

The FAW.4 was considered to be worse than the FAW.1 in this respect and during 180 degree turns at 45,000 ft when banking at 45 degrees with an entry speed of 0.90 IMN, the exit speed was 0.85 IMN. If the bank angle was increased to 60 degrees at the same entry speed, the exit speed had fallen further to 0.815 IMN. Pilot technique also affected speed loss as any rough handling would mean even greater losses than those quoted. Speed loss, buffeting and longitudinal stick forces were all slightly worse when ventral fuel tanks were fitted but in general, manoeuvrability was only marginally impaired.

The 'all-flying' tail came into its own during high speed flight and the Javelin could go supersonic in a dive from 45,000 ft. At 0.95–0.96 IMN there was a nose-down trim change followed by a nose-up trim change at 0.98 IMN which necessitated a push force on the stick to get the aircraft to its limiting speed above 35,000 ft of 1.04 IMN. The nose-down trim change was less violent in the FAW.4 than the FAW.1 and occurred at a higher Mach number (0.95 IMN as compared with 0.93 IMN). The maximum level speed of the FAW.4 however was slightly lower than the FAW.1 as it could just reach 0.88–0.89 IMN at 48,000 ft (0.90–0.91 IMN at 45,000 ft). The high speed characteristics of the Javelin when fitted with ventral tanks was little different to those in the clean condition and top speeds attainable in level flight were the same. The limiting speed above 20,000 ft with ventral tanks however was 0.95 IMN which was easily attainable in a dive.

The Javelin FAW.4 was also inferior to the FAW.1 during acceleration tests in level flight as it took 1 minute 22 seconds to accelerate from 0.85 to 0.90 IMN at 45,000 ft which was 28 seconds longer than the FAW.1. With ventral tanks fitted acceleration was significantly worse. As any interception with a cannon-armed Javelin was likely to result in some form of tailchase, this was a particular concern. However if the FAW.4 was above its target this could be used to advantage as speed could be picked up quickly in a dive. At

45,000 ft a dive of 2,000 ft was sufficient to increase speed from 0.85 IMN to 0.90 IMN.

The Javelin FAW.4 could slow down extremely quickly and its performance in this respect was just as impressive as the FAW.1 thanks to its massive airbrakes. Once again these did not open fully until airspeed was below 430 knots but once this figure had been reached deceleration was extremely rapid. Above 15,000 ft a good rate of deceleration could be obtained while still at full power and it was recommended that this technique be used for speed control during interceptions. At low level however the violent nature of the deceleration could be an embarrassment and possibly hazardous at night. The FAW.4's engine handling was praised as the throttles could be used roughly at all heights without fear of compressor stall or flame out. On many sorties during the trial the aircraft were flown at full throttle for periods of up to forty-five minutes without exceeding the engine limitations. All attempts at relighting were successful and single engine flying presented no problems as the slight asymmetric effect was easily trimmed out by the rudder down to speeds as low as 140 kts IAS at full power.

Flying on instruments was easy enough on the Javelin FAW.4 at medium speeds and altitudes but became increasingly difficult at higher values because of the almost constant need to re-trim to fly the aircraft accurately. This was especially so at speeds above 0.95 IMN where the lack of an autopilot was most felt. At approach and circuit speeds the aircraft was easy to fly, however at low speeds in turbulent conditions it was more difficult due to the need to apply coarse control movements to maintain an accurate heading and glide path. A QGH/GCA procedure was normally flown at 240 kts IAS with 6,750 rpm (7,000 rpm with ventral tanks) and airbrakes in the fully extended position.

Final approach and landing were straightforward with little change of trim when the flaps and undercarriage were lowered. The FAW.4 touched down normally with no tendency to float, a characteristic that had been noted on the

FAW.1. The average landing run was about 1,600 yards although this could be reduced to 1,000 yards with heavy braking. Crosswind landings were carried out successfully in conditions of 30 knots at 90 degrees to the runway, the normal technique being to fly the aircraft onto the runway to prevent any tendency to float. Once on the ground the Javelin had a pronounced tendency to weathercock into wind but on a dry runway this was readily controlled with rudder and brake. Landing with a crosswind on a wet runway was rather more difficult and called for careful handling. Overshooting did not require full power and if fuel was short a power setting of 7,500 rpm was sufficient. It was best to raise the flaps before the undercarriage as this avoided any trim change.

When landing in heavy rain an added difficulty was very much reduced forwards visibility, even when rain repellent had been applied on the windscreen. In extreme conditions a direct vision panel could be opened but with the panel open at airspeeds above 180 knots the noise level in the cockpit was high and air blast through the opening was uncomfortable. The general noise level was such that it was likely to make radio communications difficult. In fact Pilot's Notes went so far as to state the following 'Landings in heavy rain should be avoided as the forward view is bad and little assistance can be gained from opening the DV panel unless a curved approach is made'. This was a somewhat surprising admission considering that the Javelin was supposed to be an all-weather fighter.

Pressurisation and heating for both cockpits were provided automatically when the cockpit was closed and temperature could be adjusted by the pilot. The windscreen and canopy were kept clear by a warm air system which generally worked well, although during prolonged descents with the throttles at idle, vision could be lost through misting or icing. This could easily be cleared however by increasing rpm and pilots no longer had to scrape ice from the insides of their windscreens or use a glycol-soaked rag to see where they were going.

As the principle task of the Javelin was to intercept bombers armed with nuclear weapons the aircraft's radius of action was an important aspect of its overall performance. As the FAW.4 could carry ventral fuel tanks its endurance was much greater than the FAW.1 and sortie times in the order of 1 hour 30 minutes were easily achieved. This compared to around 45–50 minutes for the FAW.1 which relied on internal fuel only. The use of ventral tanks however adversely affected performance and although a Javelin FAW.4 could intercept a target at a radius of action of 330 nm, the fastest target it could take on was one flying at 0.77 TMN at 45,000 ft. This meant that when intercepting high speed targets (0.85 TMN+ at 45,000 ft) the aircraft had to be flown in the clean configuration and a much reduced radius of action had to be accepted (135 nm with full bad weather reserves in this case, a figure which assumed no enemy ECM and good GCI). When intercepting targets below 35,000 ft the choice as to whether or not to use ventral tanks did not arise as the difference between the clean and tanked aircraft was extremely small. With ventral tanks fitted the radius of action at 30,000 ft (with bad weather reserves) was 253 nm.

A brief weapons evaluation was conducted during the trial in which it was found that the FAW.4 could carry out quarter attacks up to 35,000 ft against targets flying at 0.70 TMN. At greater altitudes all curve-of-pursuit attacks ended as stern attacks regardless of target speed. The 'all-flying' tail of the FAW.4 gave some improvement in target tracking at high Mach numbers and satisfactory cine gun results were obtained up to 0.90 IMN in dives above 40,000 ft. The fact that the Javelin was likely to be involved in stern attacks was considered to be its most serious limitation, especially when operating against bombers armed with comparable tail armament. The performance of the AI.17 radar was evaluated and during interceptions of other Javelins at

45,000 ft the average detection range was 11 nm with AI lock-on range at 3 nm.

Most of the interception work with the Javelin FAW.4 was carried out at high altitudes against high speed targets. The standard turn used against a non-evading target was a 45 degree turn, as although there was likely to be more speed loss when compared to a 30 degree turn, the turn radii was only 4 nm as opposed to 7nm. Flying at its maximum level speed of 0.90 IMN the FAW.4 had only a 19 knot advantage over a target flying at 0.85 TMN. As speed was invariably lost during turns it was likely that the Javelin would end up behind its target but with a speed disadvantage. As acceleration in level flight was poor, closing to gun-firing range could well take time or in some cases result in failure to intercept.

When intercepting high altitude targets it was best to obtain height advantage if possible up to a maximum of 5,000 ft as anything more was liable to prejudice AI acquisition, however in some cases it was worthwhile losing altitude to gain speed. The technique which was evolved called for all turns to be made when level until the Javelin was behind the target. Up to 2,000 ft could then be lost to increase speed to 0.90 IMN and then this level was maintained until range was around 1,000 yards with the target 30/40 degrees above. A climb was then initiated to bring the target to 5 degrees above by which time range should have reduced to 500 yards.

To reduce the time to intercept the Javelin crew had to work closely with the GCI station and a technique was formulated in which the GCI controller positioned the fighter so that the target crossed ahead at 4 nm (the range equal to the radius of a standard 45 degree banked turn). The AI contact was normally at about 35 degrees angle off which closed to 15 degrees off at 4.2 nm range at which point the turn was initiated. The turn could then be eased or tightened so that the fighter rolled out behind the target at the desired range of one mile. This was called a 90 degree tangential interception.

Another type of interception was the displaced parallel head-on where the fighter was positioned so that there was a 17 nm displacement between its track and that of the bomber. The GCI controller then ordered the fighter to start a 90 degree turn towards the target when separation between them was 19 nm. On rolling out of this turn the fighter then flew straight for one minute during which AI contact was established and the target identified. From this point the interception was the same as the 90 degree tangential as described above.

Where the above displacement was impossible for the GCI controller to set up, a parallel head-on interception was attempted however after much experimentation it was discovered that the nearest to a true head-on interception that was practicable was a parallel head-on with a displacement of 8 nm which was the diameter of a turn through 180 degrees at 45 degrees angle of bank at 0.90 IMN. The range to turn was 11 nm (equivalent to an initial contact range on AI of 14 nm at 35 degrees angle off) which was within the normal detection range of AI.17. Once again the bank angle in the turn was 45 degrees and the average time from commencement of the turn to being 500 yards behind the bomber was 4 minutes 50 seconds. This type of interception produced the quickest 'kill', and therefore the least bomber penetration distance, but it was not as reliable as the displaced parallel head-on technique [the interceptions described above assume a non-evading target].

During these simulated interceptions the Javelin FAW.4's AI.17 radar came in for some criticism and AWDS went so far as to say that above 40,000 ft any success in intercepting high speed targets was 'in spite of, rather than because of, the performance of AI.17'. The range (or lack of it) put pressure on the GCI controller to be precise in his positioning of the fighter and to know exactly when to order it to turn. AI.17 also had a relatively narrow scan in azimuth which reduced still further the errors that could be accepted in the initial positioning of the fighter for its interception.

At the time of the trial the roll stabilisation of the radar was so ineffective that it was not used except in the gentlest of turns.

Having fully evaluated the Javelin FAW.4 the AWDS concluded that it was capable of intercepting a target flying at 0.85 TMN at 45,000 ft but only when flown without ventral tanks. Even with the increased performance of the Javelin compared with the Meteor and Venom night-fighters that it was replacing it was considered that efficient early warning would be needed if a 'kill' was to be made at this altitude beyond thirty miles from the coast. The situation was rather better as far as the Javelin was concerned should the target aircraft be flying at 0.85 TMN at 40,000 ft as it could be flown with ventral tanks without any significant loss of performance and radius of action was much improved.

Manoeuvrability up to 35,000 ft was adequate for a bomber-destroyer but the ability to manoeuvre above 40,000 ft decreased rapidly as speed was likely to be lost in turns. The stall and low speed audio warnings were however a big improvement. Although the flying controls of the FAW.4 were not as light and pleasant as those of the FAW.1, they were considered to be satisfactory at low and medium levels. Above 40,000 ft, where the aircraft was expected to perform its major role, handling deteriorated and instrument flying was extremely tiring. During the course of the trial XA763, which was being used for ventral tank performance checks, was damaged Cat.3 on 9 February 1957 when failure of the starboard undercarriage up-lock led to a two-wheel landing.

The Javelin FAW.4 went on to fly with Nos. 3, 11, 23, 41, 72, 87, 96 and 141 Squadrons. It was also the first Javelin (and the first RAF fighter) to undertake Exercise Fabulous where fully armed aircraft were plugged in and ready to go at a moments notice with the crews already in the cockpit. This particular duty was usually taken on by each squadron for a week at a time and had previously been undertaken solely by USAF aircraft.

CHAPTER 6

Exercise Vigilant

The first real operational test for the Javelin took place during Exercise Vigilant which was the main air defence exercise of 1957. This was a three-day exercise commencing on 25 May, the intention of which was to test the UK's reaction to raids by friendly aircraft simulating the flight profiles that Soviet nuclear-armed bombers were likely to make during the first attacks of any future conflict. Interceptions by day were carried out by Hawker Hunters and the night effort was split between night-fighter Meteors and Venoms and three squadrons of Javelins. No. 46 Squadron with FAW.1s divided its effort with W/C H.E. White and eight crews of 'B' Flight operating from Leeming with the squadron's 'A' Flight remaining at Odiham. The FAW.4s of 23 Squadron commanded by W/C A.N. Davis and No. 141 Squadron under W/C P.L. Chilton (both from Coltishall) flew from Horsham St Faith due to runway repairs at their home base. The 'attackers' were many and varied and comprised Valiants and Canberras of Bomber Command, together with Sea Hawks, Gannets, Wyverns and Vampires provided by the Fleet Air Arm. The USAF helped out with B-45 Tornados, F-100 Super Sabres and B-57s (American-built Canberras), the French Air Force provided Mystere IVs and there were sundry others ranging from the venerable Meteor, to RCAF CF-100s and US Navy A-3D Skywarriors. These aircraft were to fly just over 3,000 sorties over the three-day period of the exercise and provided a formidable test for the defending fighters. Most of the aerial activity was by day with 400–500 aircraft being launched in a series 'hammer blows' with the numbers sent against the UK being somewhat fewer by night.

In contrast to previous air defence exercises, Vigilant concentrated the attacking forces in a series of short intensive raids that presented the defences with a repeated simulation of the conditions that were likely to happen in the initial stages of an enemy attack in a global nuclear war. The night attacks had to be carefully timed due to the short amount of darkness and the raiding aircraft had to arrive at a 'timeline' approximately 100 nautical miles from the coast at 0030Z. Total night sorties by the defending fighters amounted to 426, the vast majority of these being flown by the Meteors and Venoms, with Javelins contributing sixty-seven. Total 'hostile' flights were 354. As this was the first major exercise in which the Javelin had taken part, its performance in comparison with the previous generation of night-fighters was eagerly anticipated.

The Javelin force was used against six different types of aircraft and the table below shows the heights at which these operated

Type	Valiant	Canberra	A-3D	CF-100	B-45	Meteor
Height	38–41,000 ft	40–48,000 ft	41–44,000 ft	35–40,000 ft	38–39,000 ft	34–36,000 ft

Of the sixty-seven sorties flown by Javelins, twelve were inconclusive due to failures of control with the scramble call being made too early on six occasions and too late on six more. A further fifteen sorties failed to obtain even AI contact with the attackers. About half of these were due to technical malfunction, including five cases of R/T failure and three of AI radar failure. The other abortive sorties could be attributed to failures in the ground control system as in six cases the Javelins were vectored onto friendly fighters instead of the hostile raid. A further Javelin crew found that the radio broadcasts by control were unreadable.

The remaining forty scrambles each resulted in at least one AI contact with a raid and eight Javelin crews obtained more than one contact, leading to fifty-one contacts in total. However seven of these 'first time' contacts were not made until after crossing the coast due to a variety of equipment and control failures and as a result these interceptions were classed as unsuccessful. The positions of two further contacts could not be accurately established, but in both cases the contacts were subsequently lost by AI malfunctions during pursuit.

This left just thirty-one pre-coast contacts which could have produced pre-coast interceptions. In fact three contacts were not converted until after the coast was crossed due to the Javelin pilot being committed to a long tail chase and a further three contacts were not converted at all. The remaining twenty-five contacts were all successfully converted to a 'kill' over the sea. If multiple interceptions were considered it was found that a total of thirty-eight pre-coast contacts resulted in thirty pre-coast 'kills'. However, at least two targets were known to have been successfully intercepted twice by Javelins at different points along their tracks.

The difficulty of each interception depended to a large extent on the aircraft the Javelins were up against with the Meteor and B-45 Tornado being the easiest as they flew lower and slower. The B-45 made life particularly easy for the Javelin crews as they flew with their navigation lights illuminated. The Valiants, Canberras, A-3Ds and CF-100s were rather more difficult but against these aircraft pre-coast contacts were made on 48 per cent of forty-six scrambles and the kill rate was 37 per cent. One highlight of night defensive operations was the contact that was achieved by a Javelin equipped with long-range tanks on a Canberra that was 150 miles from the coast and flying at 46,000 ft. A 'kill' was obtained while the target was still 110 miles from the coast. There were another four AI contacts at a distance of 100 miles or more from the coast, however the average detection range was just over sixty miles. One disappointing aspect of the exercise was the large distance that was needed to convert an AI contact into a kill. More than half of these took place only thirty miles from the coast. Although the Javelin had a higher speed than the night-fighters it replaced, this was matched to some extent by increased target speed resulting in only small improvements in the overtake speed.

During Exercise Vigilant a somewhat unfortunate series of incidents occurred on 25 May at Horsham St Faith and involved one of 23 Squadron's Javelins (XA732). It all began when two Hunters of 74 Squadron had different emergencies within a few minutes of each other. One had been scrambled previously but suffered engine surge at 36,000 ft so that its Rolls-Royce Avon had to be shut down and relit at a lower altitude. This Hunter arrived back at Horsham St Faith just as another was going off the end of the runway having abandoned its take off. As all this was going on a Javelin was scrambled but, unbeknown to the crew, the front fixing of one of the ventral fuel tanks failed so that the front of the tank trailed along the ground as the Javelin taxied to the end of the runway. A hole was soon worn in the tank and a spark ignited the fuel it contained, an event that immediately caught the attention of those in Air Traffic Control. Frantic calls were made but due to a mix up in call-signs the Javelin crew thought that the reference to 'fire' was for another aircraft.

In the meantime the pilot of the Hunter that had had the engine problem at altitude was about to land and having also misheard the call from ATC thought that his aircraft was on fire. He immediately shut down the engine and forced landed on the airfield, the impact being severe enough to break the fuselage behind the cockpit and fire the ejector seat, depositing the unfortunate pilot, still in his seat, onto the taxiway. He escaped with a broken elbow, cuts and bruises and shock. Although this had all been rather entertaining for the Javelin crew they now became aware of a distinct orange glow that appeared to be following them and they finally realised that it was their aircraft that was on fire. As there was no possibility of vacating the Javelin to the rear because of the flames, the only way the crew could escape was to jump from the nose from a height of 12 ft. The only injury sustained was to the pilot's feet through 'landing in the running position', he and his navigator then quickly putting as much distance between themselves and their stricken aircraft which eventually burnt itself out. The day's events soon became common knowledge at every RAF station and caused much amusement to those not involved.

Although the results for the Javelin force in Exercise Vigilant appeared to be rather mixed they were still significantly superior to the Meteors and Venoms that flew the bulk of the night-fighter sorties. Out of 359 flights their pre-coast interceptions amounted to only forty-nine. The Javelins showed marked superiority in converting AI contacts to kills, the figures for Javelins being around 80 per cent whereas for the Meteors and Venoms it was 50 per cent. This figure was even lower as height was gained and fell to 30 per cent above 40,000 ft. Variation was however noted amongst the Javelin force with the FAW.1 not performing as well as those squadrons equipped with the FAW.4 and FAW.5. This was mainly due to the fact that the FAW.1 did not have long-range tanks and sorties times were generally around forty-five minutes. The later Javelins were able to stay in the air longer and flight duration was of the order of sixty minutes, although four aircraft did manage to stay airborne for just over ninety minutes. As a result of this better endurance most, but not all, of the very long-range interceptions were by the later marks.

What pleased the RAF more than anything else in Exercise Vigilant was the fact that the Javelin managed to intercept twice as many targets before the coast when compared with the previous best night-fighter results. The ratio of improvement became even more impressive when a comparison was made with the last annual exercise (Stronghold) which was the only one in which raiders' height and speed was in any way comparable with Vigilant. The increased operational ceiling of the Javelin brought about a marked improvement over the achievements of previous exercises. In the past there had been a preponderance of targets above the defending fighters on AI contact and some raiders were above the effective ceiling of the night fighters participating. This situation was reversed with the Javelin during Vigilant and about 67 per cent of targets were at or below the height of the fighter when AI contact was made.

Despite a number of failures of equipment in the Javelin itself, together with various failures of the ground control system which led to unsuccessful sorties, Exercise Vigilant showed that the new Gloster all-weather fighter had considerable promise and was a quantum leap over the Meteors and Venoms that had served the RAF well but were now quickly becoming obsolescent. In its summing up of Vigilant the RAF was of the opinion that the Javelin gave the prospect of providing the UK with night-fighter defence that was more effective than any seen since the Second World War. With further improvements to the Javelin in the form of more advanced marks with greater capability (including air-to-air missiles) it was also hoped that the prospect of being confronted by a fully effective Javelin force might well impose a degree of uncertainty on an enemy's plans sufficient as to make a first strike by bombers equipped with nuclear weapons unacceptable.

CHAPTER 7

The Javelin FAW.5 and FAW.6

One of the main criticisms of the early Javelins was a lack of internal fuel capacity which resulted in sortie times of only 45–50 minutes and a limited radius of action. In attempting to address this problem Gloster first produced the FAW.5 which carried an extra 125 gallons of fuel in each wing. In nearly all other respects it was similar to the FAW.4. The Javelin FAW.5 was tested by the AWDS and two aircraft (XA642 and XA648) were collected from Moreton Valence on 23 March 1957 with the trial beginning four days later. By 5 June a total of 110 hours flying had been carried out and testing of the FAW.5 in the clean configuration was almost complete. Although it was not directly concerned with the trial, the two aircraft involved flew fourteen sorties during Exercise Vigilant which provided useful information on the employment of the Javelin FAW.5 with ventral tanks.

The increased internal fuel capacity on the Javelin FAW.5 was achieved by rearranging the positions and shapes of the five tanks in each wing so that 6,880 lbs of fuel could be carried (5,450 lbs on the FAW.4). On the FAW.5 it was possible to isolate the No. 3 tank before refuelling which reduced the fuel load by 1,800 lbs. This meant that the FAW.5 could be operated with four different fuel loads – (a) Clean aircraft with all tanks full – 6,880 lbs; (b) Clean aircraft with No. 3 tanks empty – 5,080 lbs; (c) Ventral tanks fitted, all internal tanks full – 10,640 lbs; (d) Ventral tanks fitted, No. 3 tank empty – 8,840 lbs. During the trial there was a temporary restriction on the use of ventral tanks for a time which meant that certain aspects of performance could not be tested until a later date when the restriction had been lifted. All flying during the trial was carried out with a full load of ammunition.

The Javelin FAW.5 used the same Mark 14 engine starter cartridges as the FAW.4 but unlike previous experience with this cartridge there were very few starting failures. There was no obvious reason for this other than the cartridges being from a better batch or that servicing personnel had gained more experience with this type of engine starting and were now ensuring that the starter breeches were clean and that the electrical contact was good. As with previous Javelins considerable power was needed to get the FAW.5 moving but care had to be taken to avoid damage to ground equipment when taxying in a confined space. With full internal fuel the take off run was slightly longer than the FAW.4 but was still well under 1,000 yards. Typical examples were 710 yards and 675 yards with respective headwind components of 10 and 15 knots. It was felt that until experience had been built up with the footbrakes, initial directional control was more easily obtained if the brakes were released at 6,000 rpm rather than at full power.

The handling characteristics of the Javelin FAW.5 were very similar to the FAW.4 and at the beginning of a sortie performance and manoeuvrability were either slightly better or slightly worse than the FAW.4 depending on whether the FAW.5 was being flown with empty or full No. 3 fuel tanks. Climb performance was tested with the No. 3 tanks empty and also with full internal fuel and all climbs were flown at

400 kts IAS until 0.85 IMN was reached. The FAW.5 showed a reasonably brisk rate of climb up to 40,000 ft which was reached in nine minutes with the No. 3 tanks empty (5,080 lbs fuel at take off). Thereafter the climb rate fell away progressively and 45,000 ft was reached in 12 minutes 9 seconds with the aircraft taking 16 minutes 21 seconds to reach 48,000 ft. When full internal fuel (6,880 lbs at take off) was used these times were extended so that 40,000 ft was reached in 9 minutes 45 seconds. The times for 45,000 ft and 48,000 ft were 13 minutes 35 seconds and 18 minutes 48 seconds respectively.

During the trial it became apparent that at altitudes above 30,000 ft the adverse effect of turning on the climb rate was far more noticeable on the FAW.5 than it had been with the FAW.4. The times quoted above were for straight climbs but in a situation where a GCI controller was obliged to order several changes of direction during the climb to carry out a high altitude interception the time taken to reach 45,000 ft was likely to be increased by many minutes.

After reaching 45,000 ft, and when operating with full internal fuel, it took approximately thirty minutes before the fuel load was equivalent to that of the FAW.4. This had an effect on high altitude manoeuvrability. Although all Javelin variants lost a certain amount of speed during a turn, the FAW.5's exit speed after completing a 180 degree turn was slightly lower than was the case with the FAW.4. There were also several occasions where a pilot was unable to hold a 60 degree banked turn at altitude as the audio stall warning sounded at about 55 degrees of bank. It was also extremely difficult to make accurate turns at 45,000 ft, particularly when the aircraft was heavy.

At 45,000 ft the handling characteristics of the FAW.5 up to the limiting Mach number of 1.04 were virtually identical to the FAW.4. The maximum level speed was also the same at 0.90 IMN. After climbing to high altitude however the FAW.5 normally still had more fuel than the FAW.4 and its increased weight meant that it had quicker acceleration in a dive. As was to be expected the extra weight counted against the FAW.5 during level accelerations and it required 1 minute 33 seconds to accelerate from 0.85 IMN to 0.90 IMN, as opposed to 1 minute 22 seconds for the FAW.4.

Instrument letdowns were initially flown at a power setting of 6,750 rpm and an indicated airspeed of 240 knots but once the aircraft was in the GCA pattern it was recommended that power be set to 7,000 rpm which resulted in a glide path approach speed, with full flap, of 140 kts IAS. Glide path alterations could then be made by adjusting the position of the airbrakes rather than alterations to the throttles. For a single-engined GCA using full flap, the recommended power setting was 7,750 rpm but here large power adjustments were required if the aircraft was allowed to get below the glide path.

One embarrassing aspect of the trial was the poor performance of the VHF R/T which was subject to complete fading when R/T range exceeded 100 nm and also when the aircraft was directly over the ground transmitter. Nearly 50 per cent of the sorties flown during Exercise Vigilant were abortive because the crew were unable to receive transmissions from the GCI stations. This situation seriously limited the effectiveness of the FAW.5 at the time, as its improved radius of action could not be fully exploited due to the limited range of its VHF radios [the VHF radios were eventually replaced by UHF sets but it would take several years before all Javelins were modified].

The increased internal fuel capacity of the Javelin FAW.5 meant that standing patrols became a practicable proposition. The speed selected for such a patrol in the clean aircraft had to be a compromise between the best endurance speed and the time required to accelerate to maximum speed (0.90 IMN). Pilots reported no difficulty in carrying out patrols at 45,000 ft and if no targets were immediately apparent, height could be increased slowly to more than 46,000 ft as fuel was burnt off and weight reduced. This allowed the possibility of being able to increase speed in a dive should this be necessary to engage a target. Two patrol speeds were investigated, 0.81 IMN and 0.85 IMN.

At 0.81 IMN at 45,000 ft the indicated airspeed was 215–220 knots which was adequate for performing gentle turns at this altitude. When speed had to be increased to attack a target the best technique was to initiate the acceleration by losing 3–400 ft with the FAW.5 taking a little over three minutes to reach 0.90 IMN when heavy with fuel and around two minutes when the fuel load was reduced towards the end of the patrol. Flying 150 nm from its base with an allowance of ten minutes for combat, a speed of 0.81 IMN gave the FAW.5 a good weather patrol time of forty-two minutes which was reduced to twenty minutes if a full bad weather recovery had to be made on returning to base.

If a patrol was flown at 0.85 IMN the time taken to accelerate to 0.90 IMN at the beginning of the patrol was about two minutes reducing to 1 minute 20 seconds by the end of the patrol, however the use of the higher speed reduced patrol time to thirty-seven minutes in good weather and seventeen minutes in bad weather. As flying at 0.81 IMN gave a 13½ per cent increase in patrol time in good weather it was recommended that this speed be used should standing patrols be deemed necessary. Although standing patrols were well within the limits of the aircraft's performance and the coverage of UK-based radars, the patrol line was likely to be at the limit of effective two-way VHF R/T communications [fading of the VHF radio signal when operating well away from the ground transmitter had also been experienced with the Javelin FAW.1 and FAW.4].

In spite of the fact that the all-up weight of the FAW.5 was greater than any previous mark of Javelin its performance at altitude in the clean configuration was only marginally inferior so the basic interception profiles in the horizontal plane required no alteration. However its reduced rate of climb did impact on the scramble decision as it took just over a minute longer to reach 45,000 ft and nearly 2½ minutes longer to reach 48,000 ft. Assuming that the Javelin FAW.5

was operating from a base 20 nm inland the minimum early warning needed for a 'kill' to be obtained at 30 nm from the coast was 249 nm. This was needed to allow for the fighter to become airborne, climb to 45,000 ft and then accelerate to 0.90 IMN. It would then need to turn 90 degrees towards the target before turning a further 90 degrees to roll out behind, whereupon it would close to a range of 500 yards before opening fire. This whole process took approximately twenty-four minutes, hence the need for as much early warning as possible.

In their summing up, AWDS pilots were of the opinion that there was no significant difference in interception capability between the 'clean' FAW.5 and the FAW.4, but that the increased internal fuel of the FAW.5 was a positive step forward that opened up the real prospect of a Javelin being able to attempt more than one interception per sortie. Radius of action was more than could be accepted by the coverage of UK-based ground radars and the reliability of VHF radios and at 45,000 ft and with bad weather fuel allowances the radius of action was 227 nm which extended to 297 nm in good weather.

Although the amount of time that the trials aircraft could be flown with ventral tanks was extremely limited owing to the temporary restriction on their fitment, the basic level of performance was assessed. With full internal fuel, ventral tanks and a full load of ammunition the all-up weight of the Javelin FAW.5 was increased to 39,565 lbs. The take off run was noticeably longer at 1,000 yards into a 15 knot head-wind and the amount of time taken to accelerate to climbing speed was also longer. In conditions of strong ground level turbulence instrument flying was most unpleasant and difficult.

After take off the best climbing speed of 400 kts IAS/ 0.85 IMN was again used but with ventral tanks the rate of climb was considerably less than in the clean configuration. An altitude of 40,000 ft was reached in 13 minutes 46 seconds which was four minutes longer than when ventral tanks were not fitted and 45,000 ft took a full 20 minutes 54 seconds as opposed to 13 minutes 35 seconds. Indeed the aircraft struggled to make 45,000 ft and speed had to be reduced slightly to coax it to this altitude. This meant that acceleration to intercept speed, which was already noticeably worse, took even longer. Manoeuvrability was also significantly affected by the extra weight and drag of the ventral tanks and at 45,000 ft with a maximum fuel load, a level 60 degree banked turn could not be sustained as the stall warning device was activated. Although 45 degree banked turns could be made to change heading by 180 degrees, with an entry speed of 0.88 IMN, which was the maximum achievable at 45,000 ft, the exit speed was so low (0.805 IMN) that it then took seven minutes to regain the entry speed.

Despite the limited amount of flying that was possible with ventral tanks it did not take long to discover that acceleration rates with the FAW.5 were much worse than the FAW.4 in the same configuration. Furthermore the difference between a clean and tanked FAW.5 was much greater than between a clean and tanked FAW.4. For example, at 40,000 ft the time difference between a clean and tanked FAW.4 accelerating from 0.70 IMN to 0.90 IMN was forty-one seconds, whereas for the FAW.5 the difference was 1 minute 55 seconds. If speed was allowed to fall to 0.75 IMN at 45,000 ft early in the sortie, the FAW.5 would not accelerate unless the aircraft was first dived for several thousands of feet. Using this technique the amount of time taken to reach 0.88 IMN at 45,000 ft was 10–15 minutes.

Although there was a significant improvement in radius of action this did not mean that enemy bombers could be attacked any further from the coast due to the range restrictions of UK-based ground radars at the time. With a radius of action of 427 nm with bad weather reserves (500 nm in good weather) the Javelin FAW.5's range capability was also in excess of any projected radar system so its increased endurance could not be exploited in this respect.

With increased fuel capacity the tanked Javelin FAW.5 was capable of spending longer on a standing patrol but even here there were considerable difficulties. At 45,000 ft even the most gentle of turns at maximum power restricted speed to 0.80–0.82 IMN and the amount of time taken to accelerate to a speed of 0.88 IMN was eight minutes. It was best, therefore, not to allow speed to drop below 0.85 IMN, even if this meant that height had to be lost in the process. An alternative technique was to commence the patrol at 40,000 ft and allow the aircraft to climb slowly to 45,000 ft as fuel was consumed. A power setting of 8,000 rpm was sufficient to maintain 0.85 IMN at 40,000 ft and at this power height could be increased to 43,000 ft after 30–40 minutes and 45,000 ft after 60–70 minutes. This cruise climb patrol was less fatiguing for the pilot and was to be recommended if the threat level allowed. The length of patrol for each method was sixty-nine minutes with full bad weather reserves.

The problems that were experienced with the tanked FAW.5 at high altitude were such that it could not be considered an effective interceptor until it had been airborne for up to an hour, by which time it had reduced weight by burning off fuel. Furthermore, with full ventral tanks the Javelin FAW.5 stood little chance of intercepting high speed targets above 30,000 ft if it was scrambled after early warning by UK-based ground radars. This was attempted during Exercise Vigilant with rather embarrassing results. A further consideration for pilots was the fact that the aircraft's take off weight was considerably higher than its permitted landing weight. It took about thirty minutes of flying at 500 kts IAS at sea level before weight was within limits for landing.

The next Javelin in number sequence was the FAW.6 which differed from the FAW.5 in that the AI.17 radar had been replaced by AI.22. As with the FAW.2 this resulted in an increased all-up weight and the shorter nose was again evident. Once again the total number of FAW.6's produced was small and thirty-three were made in two batches. The first FAW.6 was XA815 which was flown for the first time on

14 December 1956 and evaluation took place up to CA release which was obtained on 15 August 1957. The CFE did not get its hands on the aircraft for another two months however when a trial was conducted to compare the FAW.6 with the FAW.5. On this occasion, instead of aircraft being delivered from the manufacturer, aircraft were borrowed from 89 Squadron which had already converted to the FAW.6 at Stradishall. The trial was carried out from 14 October to 13 November 1957 and a total of nineteen sorties were made totalling 20 hours 45 minutes flying time. All of this flying was done with a full load of ammunition but ventral tanks were not fitted.

The ground handling of the FAW.6 was similar to previous marks of Javelin and the only discernable difference in flying characteristics was that it appeared to be more stable laterally above 40,000 ft than the FAW.5. Due to the FAW.6's increased take off weight the amount of time it took to reach various altitude levels was longer than the FAW.5. It took 10 minutes 57 seconds to reach 40,000 ft (as opposed to 9 minutes 45 seconds for the FAW.5) and 45,000 ft took 15 minutes 10 seconds from the beginning of the take off roll as against 13 minutes 35 seconds for the FAW.5. A total of five different aircraft were used during the trial and it was noticed that there was some variation in top speed between them, the range being 0.88–0.91 IMN. This difference in performance with individual aircraft had also been noted during the testing of previous marks of Javelin. In most other respects the general handling and instrument flying qualities of the FAW.6 were similar to the FAW.5

Overall the performance of the Javelin FAW.6 was considered to be slightly inferior to the FAW.5. Although there was hardly any difference below 20,000 ft, climb rate began to fall away, especially above 30,000 ft. The average maximum level speed of the FAW.6 was 0.01 IMN less than that of the FAW.5 and on average it took eleven seconds longer to accelerate from 0.85 IMN to 0.90 IMN (once again considerable variation was noted among individual aircraft). The combat radius of action at 45,000 ft was also slightly inferior at 19 nm less than the FAW.5, this difference decreasing with altitude so that at 2,000 ft the radius of action was the same.

The Javelin FAW.6 first flew with 89 Squadron which was re-numbered as 85 Squadron in November 1958. It was also operated by 29 Squadron but its service life was relatively short and it had been replaced by the Javelin FAW.8 and FAW.9 by 1961.

CHAPTER 8

Spinning the Javelin

As the Gloster Javelin had a unique planform, very little was known about the behaviour it was likely to exhibit during certain critical aspects of flight, in particular the stall and spin. Tests were carried out in the vertical tunnel at RAE Farnborough with 1/40 scale models from August 1948 to May 1949 however no-one could be sure if the real thing would exhibit the same tendencies or not.

The models were made of medium density balsa wood with hardwood inserts for strengthening. Although the tests showed that the spin and recovery characteristics of the Gloster F.4/48 delta aircraft were qualitatively like those of conventional models, because of the very low damping in rotation the recovery behaviour was poor. It was considered, therefore, that the ability of the full scale aircraft to recover even from incipient (two turn) spins was doubtful. It was discovered that the model was particularly sensitive to the application of rolling moments and that pro-spin moments caused an improvement in the recovery. Some aspects of the RAE report were to be prophetic – 'The shielding effect of the wing on the empennage on such an aircraft as this must be of primary importance ... the tailplane effectiveness will also be much reduced by the effect of the wing wake on it.'

A further series of model tests were carried out at RAE but in this case very much larger models of 2/11 scale were used. Two types of model were used, one was known as the 'silhouette' model as although it had the correct wing planform it did not exhibit the body contours of the 'correct' model which was a scaled-down Javelin. The models were dropped from a captive balloon from a height of 3,500 ft and were fitted with parachutes to save them after the spinning results had been attained. The model was rotated in the desired direction of spin and then released. Its behaviour was captured by a number of cameras from the ground and the balloon.

For each of the drops the controls of the model were set for the spin, i.e. elevators up, rudder pro-spin and ailerons neutral. These control positions were maintained until recovery action was taken. The technique used for recovery was to reverse the rudder, deflect the ailerons to full pro-spin (in a normal aircraft this would involve stick to the right in a spin to the right) and then move the elevators down after three seconds. This full control was then held for a total of five seconds after reversal of the rudder, the controls then being centralized. The results suggested that the Javelin would have a slow oscillatory spin and that recovery should be satisfactory if normal recovery action was taken. If this failed, putting the ailerons in the pro-spin direction would considerably improve the chances of recovery.

The model tests had shown that the Javelin would recover from a spin if the normal technique was used however when the aircraft itself was tested it exhibited unique spinning characteristics. Much of the Javelin spin programme was flown by W/C R.F. 'Dicky' Martin who had replaced Bill Waterton as Gloster's Chief Test Pilot in March 1954. Before this could be carried out the third production Javelin FAW.1 XA546 was lost in a fatal accident on 21 October 1954. The aircraft was being flown by thirty three-year-old F/L R.J. Ross, a graduate of the Empire Test Pilots School, RAE

Farnborough and who was on loan to the manufacturers from that establishment.

The purpose of the flight was to obtain trim figures for certain engine rpm and airspeeds. In order to improve the stability of WT830 a rubber bungee had been fitted to give the control column forward bias. The amount of stability put into the aircraft by such means showed itself in the amount the tailplane angle had to be altered to trim the aircraft at different speeds. The bungee strength on WT830 (after modification) was accepted by A&AEE but during production testing it was discovered that trim settings varied between aircraft. A programme was thus commenced using XA546 to find the reason for the differences and to bring the trim settings on all Javelin aircraft into line with those of WT830.

F/L Ross took off from Moreton Valence at 0954 hrs to establish the trim figures for various airspeeds at 10,000 ft and 6,000 rpm. He called air traffic control five minutes later to say that he was over Filton at 5,000 ft and transmitted for a bearing check. This was given and acknowledged. The next message was received at 1005 hrs and read – 'over Severn, spinning, trying to recover.' About eight seconds later the pilot of a second Javelin who was also airborne heard F/L Ross call 'throttling back', although ATC logged the message as 'still spinning.' No further message was received and eye witness accounts stated that the aircraft was seen diving steeply whilst rotating about its longitudinal axis until it entered the sea. The Royal Navy was requested to assist in the salvage of the Javelin but after four weeks the operation was abandoned. The search for wreckage was continued by Gloster until March 1955 but apart from small items (including the pilot's helmet) nothing was found. The cause of the accident remained obscure but was to prompt a thorough investigation of the Javelin's stall/spin characteristics.

Although intentional spinning was prohibited on the Javelin during its RAF service, its spin characteristics were included in Pilot's Notes and made for sobering reading. The direction of the spin was quite often unpredictable and often changed from turn to turn. The rate of rotation was very slow with the nose pitching up and down fairly regularly through as much as 70 degrees and the rates of yaw and roll varied with changes in the pitching. Generally the forces on the stick were light throughout with no 'kick back' but the rudder was likely to move fiercely one way and then the other and it was recommended that the feet be kept lightly on it throughout the spin. Readings on the airspeed indicator were likely to vary from about 90 knots to 'off the clock'.

Once it was clear that the Javelin was in a spin with the control column fully back it was recommended that full aileron be applied in the direction of spin (i.e. pro-spin). Having done this the control column had to be moved fully forward into the corner while keeping the feet lightly on the rudder pedals. It was unlikely that this action would have any effect for one or even two turns and during this time it was possible that the direction of the spin would reverse. If this happened the control column had to be moved as quickly as possible over to the other side while keeping it fully forward. Throughout this action no force had to be used to oppose any rudder movement.

The way a Javelin recovered from a spin could not be guaranteed but in most cases rotation would cease and the aircraft would hang in a nose-down attitude for a second or two. The control column had to be held fully in its corner until the aircraft did a sharp nose-down pitch or 'bunt'. The manoeuvre usually produced considerable negative 'g' of around –2½g and was quite unmistakable. This was the sign that the spin had stopped. Afterwards the speed increased rapidly and only then had the controls to be centralised and the aircraft eased out of the dive. If the pilot attempted to bring the Javelin out of the spin too soon by centralising the controls after the rotation had ceased but before the 'bunt' had occurred he ran the risk of a further spin occurring with consequent delayed recovery. The rudder was best left alone as it would centralise itself once recovery was complete.

In some cases, after taking recovery action, the aircraft would enter a fast spiral in a steep diving attitude. This spiral could be in the same direction as the applied aileron, or against it but could be recognised because the pitching characteristic would have ceased and the airspeed would be increasing. Once the speed was over 200 knots the controls could be centralised and the aircraft eased out of the dive. Once again the rudder had to be left alone and could be expected to centralise on full recovery.

The Javelin's unpredictable spin characteristics were demonstrated on 8 December 1955 when S/L (later Air Vice-Marshal) David Dick of 'A' Squadron A&AEE Boscombe Down took off in FAW.1 XA561 to check manoeuvre boundaries and spinning for service clearance. Two spins were carried out from 44,000 ft which proved that the spinning characteristics were similar to those previously found during trials with XA548. Manoeuvre boundaries were then checked by setting the aircraft up into a left-hand spiral at 43,000 ft. With airspeed at 180 kts IAS and with the aircraft in the pre-stall buffet at 3g, the rudder suddenly went hard over to the right on passing 39,000 ft and the Javelin went into a right-hand spin. S/L Dick centralised the control column but the rudder then went fully over to the left and the Javelin went into a spin in that direction.

Once the spin had stabilised Dick realised that it was different from any he had experienced before as it was flat, with a high rotational speed approximately twice as fast as had occurred previously. In contrast to other spins there was also no pitching. The aircraft was fitted with an anti-spin parachute but even this failed to bring the Javelin out of the spin and Dick eventually ejected at 8,000 ft over the Isle of Wight. His aircraft continued to spin and crashed near Ashey.

Another Javelin to be lost after a spin was FAW.9 XH755 of 33 Squadron on 18 May 1962. Blue section, consisting of four Javelins, had been detailed for high level battle formation and 2 v 2 cine gun training over the North Sea. For the latter part of the exercise the limitations were a minimum height of 30,000 ft and a maximum of 4g. After splitting into pairs, one successful interception was carried out and the pairs then separated for a second attack. During this attack the pilot of XH755 (Blue 2) found himself in a good position to take cine film of Blue 4. He manoeuvred his aircraft so that he was line astern at a range of 2–300 yards by which time he was at 32,000 ft with a speed of 200 kts IAS. To maintain his position behind the other Javelin he put his aircraft into a seventy degree bank to starboard and pulled on the control column, producing an acceleration of around 1½g. He had also selected full power with reheat, as had Blue 4. A few seconds later XH755 flicked violently to port, out of control, without any audio warning or pre-stall buffet. The aircraft flicked about two complete turns before the stall warning sounded and it then began spinning to port.

At various times the Javelin was seen to be in a flat spin and at others in a nose-down attitude. The rate of rotation was also not constant. Full in-spin aileron was applied with the control column held fully forward but the controls felt sloppy and after several turns without any sign of recovery the pilot told the navigator to eject. After he had departed at a height of about 17,000 ft the pilot tried to rock the aircraft using harsh throttle and control movement, combined with flap and airbrake selections. This also had no effect and the pilot ejected as the Javelin was passing 10,000 ft. The pilot of Blue 4 saw the aircraft crash into the sea still in a flat spin, catching fire on impact. Although the eye witness accounts varied somewhat, it appears certain that the parachutes of both the pilot and navigator of XH755 did open, but only the pilot was picked up by rescue helicopter after about half an hour in the water. Despite an extensive search the navigator was not found. This accident highlighted once again the difficult spin recovery of the Javelin. On this occasion however the initial loss of control was the result of encountering wake turbulence which was exacerbated by the use reheat.

Not all spins ended in loss of the aircraft however and on 6 May 1963 Javelin FAW.9 XH884 of 11 Squadron was successfully recovered from a multi-turn spin from 30,000 ft. The pilot was the leader of a pair that had been detailed for a training exercise to include battle formation flying and mutual interceptions. Around thirty minutes after take off the pair were attacked by two *Luftwaffe* F-86 Sabres which approached from their 10 o'clock position and above. The Javelins immediately turned towards the attack and the No. 2 Javelin, who was in front at the time, pulled up and to the right, taking the lead F-86 with him. By this time the second F-86 had overtaken the No. 1 Javelin and was in front of, and inside, the latter's hard turn to port. The pilot of the Javelin then manoeuvred his aircraft so that it was 500 yards behind the F-86 in a steep turn to starboard. As the latter was slightly high and outside the Javelin's turn, the pilot decreased the angle of bank slightly to increase the radius of turn. The aircraft was then climbed to take out the height difference and with the decreased angle of bank the aircraft was just off the buffet at 200 kts IAS. The amount of 'g' that the Javelin was pulling at this point in the combat could not be established but the crew's 'g' suits were inflated.

When the pilot judged that he was in a firing position, he attempted to synchronise the radius of his turn with the radius of the F-86 by increasing the angle of bank. At this point the navigator attempted to inform his pilot that indicated airspeed was 190 knots and decreasing but he had difficulty making himself heard due to intermittent noise from the stall warning system. When the angle of bank was increased, pitch up occurred and airspeed fell away rapidly as the aircraft flicked 360 degrees to port, before entering a spin to starboard from 30,000 ft. By now the stall warners were sounding continuously and as airspeed was below

170 knots the low speed warning was adding to the noise in the cockpit. The pilot immediately took full spin recovery action by applying full pro-spin aileron and moving the control column fully forward. During the combat full power and reheat had been selected and it was only after two full spin rotations that the pilot cancelled reheat and throttled back. The aircraft continued to spin and as he could not communicate with the pilot because of the noise from the various warning devices the navigator (P/O A. Evans) ejected as it passed 19,000 ft, the Javelin having completed seven complete rotations by that time. The pilot kept the controls in the same positions for two more turns but the navigator's ejection made no difference to the spin characteristics. Full power and reheat were then selected again and the rate of rotation gradually began to slow. After a further two or three turns the spin stopped altogether and the aircraft hesitated before pitching forward into the distinctive bunt manoeuvre, allowing the pilot to recover from the dive at an altitude of 14,000 ft. The aircraft was then flown back to base at Geilenkirchen where a successful landing was made [P/O Evans survived but suffered injuries whilst ejecting}.

CHAPTER 9

The Javelin FAW.7

The Javelin FAW.7 was built in greater numbers than any other variant with production amounting to 142 with Gloster being responsible for 85 with the remaining 57 emerging from the Armstrong Whitworth plant at Baginton. The FAW.7 was powered by up-rated Sapphire Sa.7's of 11,000 lbs thrust which were started by an isopropyl nitrate system, otherwise known as AVPIN, and take off weight with ventral tanks was now up to 40,270 lbs. To cater for the greater airflow needed by the Sapphire Sa.7 the inlet ducts were enlarged. Armament comprised four 30 mm Aden cannons as seen on previous marks of Javelin however the FAW.7 was designed to be the first variant capable of carrying de Havilland Firestreak heat-seeking air-to-air missiles.

The crew were provided with more advanced Martin Baker Mk.3JS ejection seats which incorporated an all-in-one quick release for oxygen, R/T, anti-g and air-ventilated suit connections. Other innovations included an extended rear fuselage and angled jet-pipe nozzles (in place of the characteristic pen-nib fairing of earlier marks), a fully powered rudder with hydraulic artificial feel which was fed with pitot-static pressure from an additional pair of dorsal pitot heads and an automatic yaw damper on the rudder control circuit. The problem of forwards visibility in heavy rain was addressed with an air blast windscreen rain dispersal system and there was an improved cabin air-conditioning system incorporating shut-off cocks to isolate the cockpits from outside air when flying through contaminated atmosphere. There was now a rear view mirror for the pilot and there were numerous alterations to the layout of the cockpit and the instruments.

The Javelin FAW.7 was tested by the AWDS with flying starting on 19 March 1958. By the time that the trial was complete on 24 June a total of 164.25 hours (127 day and 34 night sorties) had been flown using four aircraft with the majority of the trials work being shared by XH747, XH749 and XH752. XH748 was also used but suffered Cat.3 damage in an accident on 2 April in which it lost a wheel on take off and as a result only contributed fifteen hours. Serviceability of the remaining aircraft was adversely affected by engine starting difficulties, in particular the unsatisfactory behaviour of the starter cartridges and a series of major defects in the starter motors. The turbo-starter as fitted to the Sapphire Sa.7 used AVPIN which was ignited by a small cartridge. Each starter had a triple breech and an AVPIN tank capable of holding two gallons which should have been sufficient for three starts but the problems that were experienced led to the trial frequently being delayed.

Assuming that the engine start system worked as it should, scramble times for the FAW.7 were similar to previous Javelins with aircraft in the air after approximately fifty seconds for a night take off from an ORP with the crew at cockpit readiness. Owing to starting difficulties however the average time was more like ten minutes. When scrambling from a lower readiness state scramble times were slightly longer than previously due to increased time needed to adjust the straps of the new Mk.3JS seat. When taking off at night in still air the Javelin FAW.7 was airborne comfortably in 700 yards however with full ventral tanks the take off roll was more like 900 yards. The extra power of the

Sapphire Sa.7 engines needed to be handled with care when manoeuvring on the ground and tyre wear was substantially reduced (and no time lost) if the throttles were opened up slowly and not above 70 per cent rpm when turning onto the runway. The brakes of the FAW.7 showed a considerable improvement over previous marks with much improved operation both in the rate of response to toe movement and in the self-centring action of the rudder pedals.

The amount of time taken from brakes off to achieving the best climb speed of 450 kts IAS was only 1 minute 10 seconds and a firm backward pull had to be made on the control column to adopt the correct climbing attitude which on a clean aircraft was about 19 degrees nose up. Best climb speed became 0.85 IMN from about 12,000 ft. The FAW.7 showed a considerable improvement in rate of climb when compared with previous marks and the clean aircraft reached 45,000 ft in 8 minutes 10 seconds which was 4½ minutes quicker than the FAW.4 and 5½ minutes better than the FAW.5. With full ventral tanks the FAW.7 got to 40,000 ft in 7 minutes 10 seconds and 45,000 ft in 10 minutes.

The Javelin FAW.7 had excellent acceleration up to 40,000 ft but from this altitude there was a gradual deterioration and by the time that 48,000 ft had been reached acceleration was so slow that it was essential that as little speed as possible was lost during turns. The FAW.7 did not decelerate as quickly as earlier (and lighter) marks of Javelin and at low level it was considered that it might be necessary to throttle back as well as selecting airbrakes to achieve the reduction in speed that was desired. When flying at low level the maximum permissible IAS of 535 knots was easily reached and with the extra power available this could be maintained even with full ventral tanks and with the engines operating at 87 per cent rpm. At high altitudes there was a small but useful increase in top speed which gave ground controllers and pilots much greater tactical freedom, especially when intercepting targets above 43,000 ft.

As with previous marks of Javelin the question as to whether or not to operate the FAW.7 with ventral tanks was a difficult one to answer as it depended very much on what height the aircraft was to be flown. The tanks themselves were large and not easy to manhandle onto and off the aircraft but for low level interceptions they were essential as the extra fuel was needed and there was no performance penalty. At high altitude however the use of ventral tanks degraded performance so that in a straight climb to 50,000 ft it was necessary for the Javelin to burn off fuel at that altitude before it was capable of intercepting a target. Once the fuel in the ventral tanks had been used however, there was hardly any performance penalty.

The improved acceleration of the FAW.7 meant that standing patrols flown with fuel efficient power settings could now be considered and the recommended technique was to climb to 40,000 ft and loiter at 215 kts IAS (0.73–0.74 IMN). Should a target appear, acceleration was good at this altitude and it was calculated that an FAW.7 with ventral tanks could patrol above 40,000 ft at 100 nm from base for one hour. If the Javelin was to be used for standing patrols well out from the coast, fuel conservation took on a new significance and several different methods were evaluated when returning to base to ensure that maximum fuel economy was achieved. The most satisfactory combination of fuel economy and ease of instrument flying involved a descent at 0.85IMN/350 kts IAS with throttles closed and full airbrakes. A descent from 40,000 ft to 2,000 ft in this manner took only 2 minutes 45 seconds and burnt only 200 lbs of fuel. A gentle pull on the control column was necessary at 4,000 ft to ensure that recovery had been completed by 2,000 ft and the same technique could be used to arrive over base for a QGH procedure.

In extreme cases, range could be extended by gliding and in a glide from 50,000 ft to 2,000 ft with both engines set to idle the amount of fuel used was only 1,750 lbs. As 250 nm was covered during the glide (which took 45 minutes) the fuel used was 7 lbs/nm. However even this figure could be improved upon and in a glide with one engine stopped and the other at idle, only 650 lbs of fuel was used. This covered 147 nm in 27 minutes which equated to 4½ lbs/nm.

A closer look at the Javelin FAW.7's handling showed that there was little need to hold the aircraft down after take off to achieve the best climb speed. As acceleration was much improved, the aircraft could be allowed to climb gently so that the climb speed of 450 kts IAS was reached at about 1,500 ft [this was particularly useful at night or when flying in poor visibility]. In the climb the fact that the nose position was much higher meant that the artificial horizon was of little use as a pitch indicator until near the top of the climb. The best instrument technique was to refer to speed, heading and to a lesser extent the rate of climb indications.

For its role, the FAW.7's manoeuvrability was considered to be adequate in the clean configuration up to 45,000 ft or up to 40,000 ft with ventral tanks, however above these altitudes it deteriorated rapidly. Compared with earlier marks the buffet in turns was slightly more pronounced and there seemed to be a wider margin between the onset of buffet and the sounding of the stall warning device. Maximum rate turns were best made inside the buffet boundary but outside the limit set by the audio warning which, when it sounded, was usually accompanied by stick force lightening and a rapid loss of speed. Accurate height control at high altitude in other than gentle turns was difficult and concentration was needed to anticipate trim changes.

During simulated high quarter attacks at 40,000 ft at speeds of 0.91–0.94 IMN when 'g' loadings were increased, slight pitch oscillations were likely to set in which could prevent steady tracking and gun aiming. This could ruin an otherwise effective attack and would continue to be a problem until the introduction of a pitch damper on later aircraft. At a speed of 0.95 IMN longitudinal control forces were too heavy for a successful attack so 'bouncing' a target from above in which speed was increased to 0.95 IMN or above gave no advantage whatsoever. In such a situation it was not until speed had been reduced by use of airbrakes that a successful attack could be made.

The Javelin FAW.7 was fitted with AI.17 radar that was identical to the FAW.1, 4 and 5. As was to be expected the radar performance was similar to that experienced in squadron service with the earlier aircraft with average detection ranges on other Javelins at high altitude of around 14 nm. Although the increased top speed of the Javelin FAW.7 was only small it was still a significant percentage of the speed advantage it had over its most likely adversary, the Tupolev Tu-16 Badger. Of greater importance however, was the FAW.7's improved acceleration which in just over half a minute enabled it to regain the speed lost in a 30 or 45 degree banked turn through 90 degrees. At these angles of bank turn radii were virtually the same as those for the Javelin FAW.4, but the FAW.7 was able to make these turns at greater altitudes for comparable speed losses.

The technique for intercepting targets when under close control did not differ from that recommended for earlier marks of Javelin. From whatever angle the Javelin approached its target, the job of the ground controller was to position the Javelin so that the target would cross ahead of its line of flight at a range that would allow it to make a banked turn of 30 or 45 degrees, with radii of 7 nm and 4 nm respectively, to roll out 1 nm behind the target. It was normal procedure for the navigator to give the order for commencement of this turn to which he would then call for variations in the rate of turn to allow for earlier positioning errors.

The improved performance of the Javelin FAW.7 enabled a good crew to accept greater positioning errors than hitherto and still bring off a successful interception. If it transpired that the target passed ahead of the Javelin at close range it was now possible to consider using maximum sustainable bank angles to position the Javelin behind the target as this reduced the turn radius substantially. Although this inevitably led to a loss of speed, the FAW.7's excellent acceleration, if used in conjunction with a dive of 2,000 ft, would quickly close range to around 1,000 yards. By climbing from this position the Javelin could get within gun firing range with the target approximately 5 degrees above.

Much of the tactical flying that was carried out during the trial involved intercepting random AI contacts without the use of close ground control. This simulated a situation in which ground control had been degraded by the use of ECM and where fighters were operating under Broadcast or Loose control, or were freelancing. In such a situation targets could be expected to appear at random anywhere within the scan pattern of the AI radar. The main variables affecting the success rate for interceptions were initial detection range and target speed but the trial showed that the FAW.7 could deal effectively with a wide variety of radar contacts. A full investigation was carried out into three types of attack where the target was flying on a 180, 135 or 90 degree collision course relative to the fighter.

During the trial the target aircraft were other Javelins and initial detection ranges averaged 14 nm, however 4 nm of this was required for the navigator to assess the contact and decide on his interception plan. As this left a range of just 10 nm the only option was to carry out a short range procedure in which the fighter made a hard turn through the target's line of flight. As radar contact was lost during the turn the fighter was blind for a time and so it was useful if the navigator could give his pilot an assessment of the target's relative height before the turn was ordered. At high altitude precision flying was called for if a successful interception was to be made and the turn had to be flown at the correct angle of bank so that the turn radius was exactly as required to position the fighter behind the target.

A short range procedure for a 180 degree collision course at 45,000 ft (i.e. where the fighter and target were approaching each other head-on) would see the fighter initiate a 60 degree banked turn to port at a range of 10 nm (a 45 degree banked turn could be used but this needed to be commenced

at 13.5 nm range). The turn would be continued through 220 degrees and the radius would be 2.75 nm (4 nm for a 45 degree banked turn). Assuming that the target did not evade, this would position the fighter so that AI contact was obtained once more and course adjustments could then be made to position the fighter behind the target at the same time as reducing range [the examples quoted here assume a target speed of 0.87 IMN and the 'fighter' Javelin operating without ventral tanks].

In the case of a 135 degree collision course where the target approached the fighter from its starboard side (in approximately the 2 o'clock position) the fighter would commence a 180 degree turn to port. Should the pilot elect to make a 60 degree banked turn the range at the commencement of the turn needed to be 4.75 miles (7.25 miles for a 45 degree banked turn). The figures for radius of turn would be the same as in the previous example and on completion of the turn the fighter would approach the target from its starboard rear quarter where AI contact would be resumed and the navigator could direct the pilot to close to gun firing range.

A 90 degree crossing contact (right to left) where the fighter and target were again on a collision course allowed the greatest degree of flexibility as turns using only 30 degrees angle of bank could be used given adequate radar warning. At the required range (dependent on the angle of bank to be used) the fighter would be put into a turn to port of 120 degrees so that it flew through the target's track and again approached from the starboard rear quarter. For a 30 degree banked turn the range at the start of the turn needed to be 4 nm reducing to 2½ nm for a 45 degree bank and 1.75 nm for a 60 degree bank angle.

The conclusions of the AWDS/CFE report were that the gun-armed 'clean' Javelin FAW.7, when scrambled from an ORP on UK early warning and given good quality GCI close control, could intercept targets flying at 0.87 TMN at 45,000 ft, 0.85 TMN at 48,000 ft or 0.83 TMN at 50,000 ft [it was to be expected that these interceptions would be made about 70 nm from the UK coast]. Performance was reduced if ventral tanks were carried and at 45,000 ft the FAW.7 was effective only against targets with a maximum speed of 0.84 TMN (0.82 TMN at 48,000 ft). The improved performance of the FAW.7 eased the navigator's task (and also that of the GCI controller) and increased the chances of success, at the same time as reducing target penetration distance.

Whenever close control was not available however the limiting factor during interceptions was the search range of the AI.17 radar. At altitudes up to 37,000 ft the Javelin was assessed as being a good gun platform up to its limiting speeds but at higher altitudes and speeds in excess of 0.91 IMN pitch oscillations could occur after pull forces generating more than $\frac{1}{2}$g. When this pitch oscillation was present accurate sight tracking or gun aiming was impossible, although it was possible for the pilot to reduce the oscillation by continued application of 'g' which reduced speed below the critical value. As the FAW.7 also had a slightly higher top speed than previous marks of Javelin it tended to fly more often at speeds at which it was affected by the nose-down trim change due to compressibility. As buffet in turns was encountered earlier, an accurate turn was more difficult to fly at altitude.

Dissatisfaction was expressed as to the Javelin's problems regarding engine starting and it was calculated that at the time of the trial there was only a one in two chance of both engines starting at the first attempt. Problems were also experienced with the fuel gauges which frequently gave in-accurate readings, a snag that had previously been reported on the Javelin FAW.5. One of the main recommendations was for the provision of a pitch damper to assist with effective gun aiming at all speeds and heights and this would eventually appear on the Javelin FAW.8.

The Javelin FAW.7 went on to equip 23, 25, 29, 33, 60 and 64 Squadrons and later aircraft carried four Firestreak air-to-air missiles as their principle armament (see Chapter Ten). On those aircraft where missiles were fitted the outer two 30 mm Aden cannon were removed. The first missile-armed FAW.7 was XH906 which was delivered to 25 Squadron at Waterbeach on 5 December 1958. A total of seventy-six Javelin FAW.7's were later brought up to FAW.8 standard with two afterburning Sa.7R engines and re-designated FAW.9.

CHAPTER 10

Javelin and Firestreak

The Firestreak infra-red homing air-to-air missile was to transform the operational effectiveness of the later marks of Javelin which were capable of carrying four such weapons on pylons, two under each wing. Development of the Firestreak (original codename Blue Jay) commenced in 1951 under prime contractor de Havilland Propellers Ltd and ground testing was completed at Larkhill in 1954. The first airborne launch took place in 1955 when a de Havilland Venom NF.2 successfully shot down a Fairey Firefly U.9 drone aircraft at Aberporth.

The missile itself was 10 ft 5 in long with a diameter of 8.75 inches and it weighed 300 lbs. The main body consisted of two main assemblies, the front section containing the guidance and control systems, the proximity fuze and the missile power supply whilst the rear section housed the motor, warhead and air bottle. Attached to the cylindrical body were four wings and four small control surfaces positioned to the rear and in line with the wings. Each of these control surfaces could move to act as elevator, rudder or aileron.

The Firestreak was designed to be used during pursuit attacks against aircraft flying at up to 55,000 ft and speeds of up to Mach 0.95 and it was normally capable of detecting the heat generated by the engines of a jet bomber at not less than 4,000 yards range. The attacking zone for the fighter was a cone of approximately 20–30 degrees astern of the target as this was where the infra-red radiation was at its greatest. Prior to reaching this position the missile had to be armed, the arming process taking around two minutes. Having acquired or detected a target, an indication was passed to the pilot who then pressed the firing button, thus completing the firing circuit. The electronic fire control equipment then selected a suitable weapon (or pair of weapons) and initiated it by running up the internal power supply, switching the homing head to lock follow on the target and igniting the rocket motor. If all four missiles on a Javelin had acquired and were fully serviceable, the order of firing was starboard outer, port outer, starboard inner, port inner. With 'pairs' selected it was not possible to launch two missiles from the same wing, so in a situation where only the starboard missiles had acquired a target, only one would fire. Firestreak also had a 'snap up' capability whereby it could engage targets flying up to 7,000 ft above the fighter.

The propulsion unit was a Magpie solid fuel motor which had a burning time of about 1.7 seconds and produced 6,500 lbs thrust. The motor was situated in the rear of the missile and was so positioned that the burning of its 61 lb charge did not alter the missile's centre of gravity. Although the burn time was extremely short the thrust developed was sufficient to accelerate the missile to about 1.6M above launch speed. Also at the rear of the missile was a bottle that contained compressed air at 3,500 lb/sq.in. This provided air to run the turbo alternator for all the missile's electrical supplies and to operate the pneumatic control actuators.

The weapon launched itself and its own guidance and control systems took it to the vicinity of the target where the warhead was detonated by a proximity fuze. In a situation where the missile actually struck the target a contact fuze caused instantaneous detonation of the warhead. If the missile failed to locate the target a self destruction system operated after a certain length of time [this was usually set at 12 seconds]. In the case where a pair of missiles was launched the port missile would be fired half a second before the starboard to avoid interaction between the two missiles.

Having already tested the gun-armed FAW.7, AWDS/ CFE turned its attention to the missile-armed version during a trial that was carried out in 1958. This should have started on 1 January 1958 but a delay in the delivery of aircraft pushed this date back to 13 June and then a lack of test equipment meant that the acquisition phase could not be commenced until 17 July. At this time the missile system was only cleared up to 40,000 ft and it was not until 19 November that clearance was received to extend testing of the system up to 48,000 ft. The trial was conducted using three Javelin FAW.7s XH755, XH756 and XH758 all with Mod.568 embodied [these aircraft were basically similar to the gun-armed FAW.7 except for the provision of the Firestreak fire control system and the removal of the two outer 30 mm Aden cannons]. Delivery was also taken of twelve pre-production missiles which were capable of acquiring targets but had dummy warheads and rocket motors, together with six drill missiles for use on performance sorties. The trial eventually amounted to 192 day and 26 night sorties totalling 205 hours 30 minutes flying time. No missiles were fired during the trial and much of the work involved assessing the Javelin/Firestreak combination up to and including target acquisition.

The trial did not go as smoothly as had been hoped as one aircraft was unavailable for five months following a starter failure and another was unserviceable for four months with recurring pressurisation, demisting and hydraulic faults. This situation was aggravated by the same start up problems that had been experienced during the trial with the gun-armed FAW.7. The aircraft were flown in both clean and ventral tank configurations with two and four missiles. In general the performance of the Firestreak-armed Javelin was inferior to that of the gun-armed version when it came to times to height and acceleration. There was however little difference in turning performance or maximum level speeds. The maximum operating height of the clean aircraft with two or four missiles was considered to be 48,000 ft. When ventral tanks were used the Javelin was only effective up to 45,000 ft although it could begin to fly higher as fuel was used up.

The major penalty that was incurred by carrying four missiles (as opposed to two) was that climb performance deteriorated above 40,000 ft. In a situation where ventral tanks were carried in addition to four missiles, the time taken to climb from 40,000 ft to 45,000 ft was increased by about two minutes. In most other respects however the performance penalties were negligible and it was considered that the operational advantages gained by the carriage of four missiles far outweighed the disadvantages. The average maximum level speed at 48,000 ft when carrying four missiles was 0.885 TMN for the clean aircraft and 0.875 TMN when flown with ventral tanks. In terms of handling the Javelin armed with Firesteaks was very similar to the gun-armed variant but at high altitudes it appeared that it did not suffer quite so much from porpoising during high speed turns.

The Javelin was also tested when carrying Firestreak missiles asymmetrically. With two missiles under the starboard wing a small swing to starboard occurred on take off with a 15 knot crosswind from that direction but this was easily held by brake and rudder. By the time that the aircraft had reached 100 kts IAS the swing could be held by rudder alone. Once airborne there was a slight yaw to starboard but this could be trimmed out. There were no handling problems up to maximum IAS and IMN although the aircraft tended to fly one wing low, even with the yaw trimmed out. The Javelin was even easy to fly in the asymmetric condition with one engine at idle. In the worst case at 150 kts IAS with the starboard engine throttled back, full power on the port engine, airbrakes out, undercarriage down and full flap, two divisions of rudder trim were sufficient to maintain balanced flight. An overshoot in this condition presented no particular problems for the pilot.

The cockpit layout of the Javelin FAW.7 was modified to cater for the fire control system for the Firestreak missiles. This included the arming switch, the operation of which started the missile cooling system and the hydraulic alternator which supplied power to the missile. The missile gyros reached operating speed after about 1½ minutes, while the electronic equipment was warmed up after approximately two minutes. At the end of the arming period the missiles were able to acquire a target and to be launched. If the missiles were subsequently disarmed a further two minutes arming period was needed. The arming and armed time was limited by the endurance of the cooling system and was fifteen minutes when two missiles were carried and thirteen and a half minutes when four were carried. There was also an arming light which illuminated when the arming switch was operated and remained on until the arming cycle was complete. An armed time clock was located in the rear cockpit to the left of the AI indicator and recorded elapsed armed time.

Acquisition lights were positioned in the pilot's cockpit to the left of the collimator glass. One light would illuminate when a single missile had acquired a target and the second light came on when acquisition had been achieved by a pair of missiles. A master armament selector switch could be used to select guns or missiles. All weapons were fired by a trigger on the control column and the master switch completed the required circuit. When the aircraft was on the ground the firing circuit was broken unless an override switch was operated in the starboard wheel bay. There was also a switch so that the missiles could be fired singly or in pairs. If 'single' was selected another missile could not be fired until 4½ seconds had elapsed and this period also applied to the firing of missiles in pairs. The missiles could be jettisoned if required by one of several methods.

A jettison firing button fired all the rocket motors at once and the missiles would self-destruct after twelve seconds. The missiles could also be ejected from their pylons by operation of a 'Clear Wings Jettison Button'. The missiles could also be jettisoned selectively by operation of the appropriate switch.

Target acquisition could be affected by many factors and the initial illumination of the cockpit acquisition light was often intermittent and could be triggered by reflections of sunlight or moonlight by various backgrounds. Steady acquisitions resulted when the aircraft was heading within 5 degrees of either the sun or moon. Cloud also affected target acquisition and in general the missile appeared to pick up the target at about the same range that it could be seen by the pilot. When the target was superimposed on a cloud background reflecting sun or moonlight, the acquisition range was considerably reduced due to the missile's inability to discriminate until the target signal became dominant. Acquisition could also be affected by aircraft contrails and in the case of a dense and persistent trail, contact was lost if this was entered by the fighter. To maintain contact therefore it was necessary for the fighter to avoid flying within the contrail. The effect of water reflection was similar to those of a level cloud layer. Reflections from sunlit patches of water sometimes caused considerable interference, particularly when the aircraft was heading towards the sun.

Whenever the missiles were pointing towards the ground when cloud cover was absent, the acquisition lights flashed with increasing frequency as height was lost, particularly when the sun was behind the aircraft, and a brightly lit coastline provided even more interference. When flying at low level at 2,000 ft AGL a different problem emerged as positive acquisitions could easily be obtained from factory chimneys. The horizon also had an effect. When tracking immediately below the horizon the acquisition lights flashed on and off, irrespective of the sun's position. Once the sight

The Gloster GA.5 prototype WD804 on an early test flight over the Gloucestershire countryside. (*Philip Jarrett*)

Another photo of WD804. It is fitted with a nose-mounted pitot boom and a fairing on top of the tailplane for the anti-spin parachute. (*Philip Jarrett*)

This view of WD804 shows the early canopy design which featured an enclosed rear cockpit incorporated two small portholes for the navigator. (*Philip Jarrett*)

WD808, the second prototype, is seen taking off from Moreton Valence on its maiden flight on 21 August 1952. (*Philip Jarrett*)

en at the SBAC Display at Farnborough in early September 1952, WD808 was displayed by oster Chief Test Pilot Bill Waterton. (*Philip Jarrett*)

e third prototype Javelin was WT827 which was flown for the first time on 7 March 1953. It is en at Farnborough later that year. (*Philip Jarrett*)

WT827 shows off the pure de[l]
wing of the early Javelins.
(*Philip Jarrett*)

Javelin FAW.1 XA560 was used
for engine trials and was later
fitted with the more powerful
Sapphire Sa.7R. (*Philip Jarrett*)

Engine start up produced
copious amounts of smoke
from the starter cartridges.
Javelin FAW.1 XA626 flew
with 46 and 87 Squadrons.
(*Philip Jarrett*)

Two Javelin FAW.1s (XA571 'C' and XA628) of 46 Squadron take off from a wet runway at Odiham. (Philip Jarrett)

Javelin FAW.1 XA627 'B' of 46 Squadron. It was later flown by 87 Squadron and became an instructional airframe (7661M) in October 1960. (Howard Fitzer)

A precise echelon formation flown by three Javelin FAW.1s of 46 Squadron. (*Philip Jarrett*)

The distinctive pen-nib fairing of the revised jetpipe is evident in this photo of Javelin FAW.4 XA631. (*Philip Jarrett*)

velin FAW.4 XA644. It was involved in a collision with Hunter XF980 on 24 August 1956, an
‑cident that killed Gloster test pilot Brian Smith. (*Philip Jarrett*)

Javelin FAW.4 XA721 was used for trials work with Gloster and at A&AEE Boscombe Down before serving with 3 Squadron. (*Philip Jarrett*)

A line up of Javelin FAW.1s of 46 Squadron at Odiham during Exercise Vigilant in May 1957. (*Philip Jarrett*)

Javelin FAW.2 XA808 is seen during service with the All-Weather Development Squadron at West Raynham. It later flew with 46 Squadron. (*via author*)

Javelin FAW.2 prototype XD158 was flown for the first time on 31 October 1955. (*Philip Jarrett*)

Four Javelin FAW.2s of 46 Squadron in formation in the summer of 1960. (*Philip Jarrett*)

Another view of Javelin FAW.4 XA644 completing the second half of a loop, a manoeuvre that was officially banned in service. (*Philip Jarrett*)

velin FAW.1 XA628 served with 46 Squadron at Odiham. (*Philip Jarrett*)

nother view of XA628, now in the markings of 87 Squadron and coded 'B'. It became an
astructional airframe in July 1961 as 7720M. (*Philip Jarrett*)

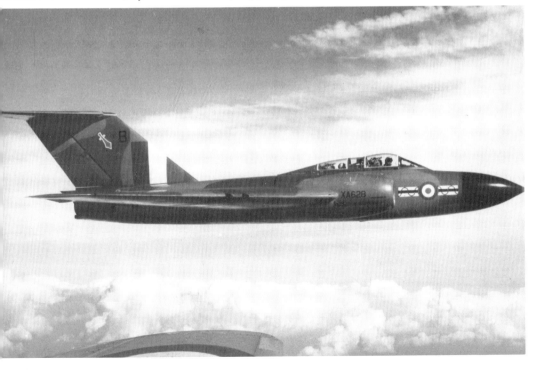

Three Javelin FAW.4s of 11 Squadron perform a break to starboard. (*Philip Jarrett*)

Javelin FAW.5 XA656 flew with 228 OCU and the All-Weather Fighter Combat School before being sold for scrap in 1964. (*Philip Jarrett*)

Javelin FAW.5 'T' in the markings of 151 Squadron. (*Philip Jarrett*)

fter flying with 89 Squadron, Javelin FAW.6 XH694 served with 85 Squadron as seen here. It was ruck off charge in 1962. (*Philip Jarrett*)

velin FAW.6 XA815 displays the colours of 89 Squadron at Stradishall. (*Philip Jarrett*)

he red and blue bands of 33 Squadron are painted on the tail of Javelin FAW.7 XH835, the personal rcraft of C.O. W/C Norman Poole. (*Philip Jarrett*)

Javelin FAW.4 XA632 was used in the preliminary trials of the de Havilland Firestreak air-to-air missile. (*Philip Jarrett*)

The prototype two-seat Javelin T.3 trainer WT841 (foreground) is seen in the air with FAW.4 XA64 (*Philip Jarrett*)

...velin FAW.7 XH758 with two Firestreak missiles on the outer wing pylons and Microcell rocket ...ds inboard. (*Philip Jarrett*)

...technician attends to a Firestreak missile during AWDS trials at West Raynham in 1958. ...*ilip Jarrett*)

Javelin FAW.9 XH965 takes on fuel from Valiant B(PR)K.1 WZ376 during in-flight refuelling trials.
(*via author*)

A fine view of the first Javelin FAW.8 XH966 which took to the air for the first time on 9 May 1958.
(*Philip Jarrett*)

was raised above the horizon however the interference ceased.

The effect of a Javelin target taking evasive action after acquisition had been made was assessed in the height band 35–45,000 ft, both by day and night. Target evasion was commenced before the fighter had reached maximum firing range at about 4,500 yards. The target was free to turn as hard as possible and increase or decrease its speed. Acquisition was never lost when the target made a level or climbing turn but if it dived during interceptions carried out by day the acquisition became intermittent because of the background effect, although it could usually be regained if range was closed. The effect of throttling back to reduce the level of radiation was tested with the target Javelin reducing power so that JPT fell back to around 400 degrees C. At a range of 4,500 yards contact was maintained and it soon became apparent that once the acquisition light became steady, it would remain on even when range was extended to beyond that of the original acquisition. The range at which acquisition was made depended on the target aircraft and its radiation signature and the height that it was being operated. For another Javelin at 40,000 ft it could be expected that acquisition could be made in excess of 4,000 yards and for an aircraft such as a V-bomber at a range in excess of 10,000 yards. When operating against another Javelin at a height of 1,500 ft acquisition could be expected at about 2,500 yards.

It was assumed that until 1962 the threat against the UK would comprise large numbers of Tu-16 Badger aircraft operating at speeds of up to Mach 0.85 in the height band 36–42,000 ft and it was also thought that this force could well be supported by IL-28 Beagles providing ECM support. At the time of the AWDS trial it was expected that any Soviet attack would most likely be in the form of a saturation raid or raids supported by heavy ECM. Such a raid might well have a frontage of some 200 nm and a depth of 60–70 nm with aircraft having a random spacing of 6–8 nm throughout the raid. Intelligence estimates at the time of the AWDS trial were that around 30 per cent of the Badger force had some form of radar jamming, and probably VHF jamming as well. By 1962 it was thought that the whole fleet would be so equipped and that more advanced radar and UHF jammers would have been introduced.

Previously it was considered that the fighter had successfully done its job if it had shot down an enemy bomber before it reached the UK coast. In the nuclear age this was no longer true and the 'kill' had to take place as far from the UK coast as possible, as attacking aircraft may well have been carrying stand-off nuclear weapons. This task was made even more difficult as the performance advantage that the Javelin possessed over the Badger was only small. It was thus advantageous for the Javelin to spend as much time as possible within the raid so that it could use the multiple-kill potential of its main and secondary armament to best effect. As part of its report on the Firestreak-armed Javelin FAW.7, AWDS also discussed the tactical aspects of its use.

The limitations of obtaining early warning from UK ground-based radars and scrambling defending fighters from cockpit readiness were such that it was considered unlikely that a Javelin FAW.7 would achieve a 'kill' at a distance of more than 80 nm from the UK coast. AWDS were also of the opinion that in this situation there would be insufficient time within the raid to obtain multiple kills. There was the possibility of additional warning time from information derived from continental radars, however accurate tracking and continuous close control might not be available. When under close control the controller had to accept that the Javelin FAW.7 armed with Firestreak was inferior to the gun-armed version in the climb and in acceleration, but it more than made up for this with increased weapon firing range and the ability of the missile to take out some 7,000 ft of height difference. This 'snap-up' capability also gave the pilot the option to lose height in the attack phase turn to maintain or increase speed, or reduce turn radius. Whenever possible it was recommended that close control be conducted with the Javelin already in the air and patrolling as extended interception ranges gave an increased chance of engaging more than one target. Interception profiles such as the displaced parallel head-on attack were still relevant with missile-armed fighters although the final positioning of the latter needed to be revised. Overall the use of missiles introduced much more flexibility and the 'height take-out' characteristic of the weapon gave more freedom to manoeuvre at altitude, as in certain cases the inevitable speed loss from steeply banked turns could now be accepted.

In any mass raid it was possible that radar cover and R/T communications might be lost due to ECM and therefore close control might not be possible. In this situation the Javelin FAW.7 would be operated in loose control mode which would require a large amount of pre-planning on the ground. Given some form of early warning from continental radars to get the Javelin force into the air, it was estimated that when the raid finally entered UK radar cover the defending fighters would be able to make contact up to 156 nm from the UK coast in still air conditions. As the Javelin's performance advantage over the Badger was small it was beneficial for the Javelin to cruise out to the inter-ception area at a higher level than the expected target height as this advantage could be used to increase speed if required.

The search range of the AI.17 radar when used against an aircraft the size of Badger was around 18 nm. Assuming a mass raid and the absence of friendly fighters it was estimated that AI.17 would show anything up to twelve targets so that the fighter would be able to attack the one that was most favourably placed. Some form of parallel head-on attack would most likely be needed so that the Javelin could be positioned to the rear and within the 20–30 degree cone behind the target. When operating under loose control (unlike close control) the choice of target was the navigator's responsibility and so in some situations it might be possible for more than one fighter to attack the same target. It was also possible for one defending fighter to intercept another. The Javelin FAW.7 was fitted with the Fighter Identification System (FIS) Mk.2 but this only provided information to an AI-equipped fighter when the latter interrogated a similarly equipped fighter from astern. Although FIS Mk.2 could be seen through a jamming 'spoke' the pilot of an attacking fighter was at a tactical disadvantage as he could not be sure that the contact was hostile or friendly before commencing his final attack phase turn.

In one sense the use of ECM was of use to the Javelin navigator as the effect was to produce vertical bands of interference on the AI radar. These spokes could be used as a means of homing on to the raid and would almost certainly be picked up at ranges well outside the normal AI search range. During a previous trial a spot frequency jammer had been used against AI.17 and the jamming spoke had been picked up at over 80 nm range. One aspect of Firestreak performance that left a lot to be desired however was the missile's limited armed time. This was a particular handicap when operating in a loose control and ECM environment as the range of the target might not be known until the weapon firing bracket was reached. In certain conditions it was possible for a Javelin to intercept a target but not be able to fire its missiles as the arming time had been exhausted.

The attack phase depended primarily on the ability of the navigator to interpret AI information to enable the Javelin to home onto a target to obtain a suitable weapon firing position and establish range, before reaching the minimum launch range of Firestreak. This was not likely to present any undue difficulty when jamming was not present and even when operating against ECM by day it was a relatively simple task to assess range by relating the target's wingspan to the gunsight. It was a completely different matter when flying at night with ECM and here it was necessary for the navigator to be able to discriminate between jamming spokes at ranges in excess of 20 nm.

The Firestreak missile was to be the Javelin's principal armament until the aircraft's retirement in April 1968. Considering that it was the RAF's first operational air-to-air missile it was an extremely reliable weapon that achieved a success rate of approximately eighty-five per cent when fired within its operating parameters. In the face of a rapidly changing threat level in the late 1950s and early 1960s it gave the Javelin force a significant boost to its operational effectiveness and allowed the Javelin to continue at the forefront of the UK's air defences until its eventual replacement by the English Electric Lightning.

CHAPTER 11

Ultimate Javelins

Many pilots considered the gun-armed FAW.7 to be the best Javelin of them all. Its Sapphire Sa.7 engines of 11,000 lbs thrust gave it a thrust increment of almost 40 per cent over the early marks of Javelin and endowed it with a much improved rate of climb and a higher service ceiling. However the adoption of the Firestreak air-to-air missile and its associated systems inevitably increased weight and drag and this had a detrimental effect on the FAW.7's performance, especially at high altitude. As the Javelin only had a small performance advantage over Soviet jet bombers, any improvement in the height band 35–45,000 ft could be crucial when it came to destroying hostile aircraft before they reached the UK coast.

In terms of Javelin development the ultimate variant was the FAW.8 which was flown by Gloster Chief Test Pilot Dicky Martin for the first time on 9 May 1958. Like the FAW.2 and FAW.6, the radar fitted to the FAW.8 was the AI.22. The aircraft was powered by Sapphire Sa.7R engines that offered a limited reheat which boosted thrust by 12 per cent to 12,300 lbs. The system was fairly rudimentary as it only became fully effective at altitudes above 20,000 ft and if used at low level, power was actually reduced. This was because at lower levels the fuel pumps could not deliver enough fuel to supply the engines at maximum thrust and the reheat system as well. Thus if reheat was selected at low level, the amount of thrust produced overall was less than that when flying with the throttles set for maximum dry thrust. When flying above 20,000 ft the fuel pumps were able to deliver sufficient fuel for the engines and the reheat system and this restored most of the high altitude performance that had been lost due to carrying four Firestreak missiles under the wings.

When compared with previous Javelins, the FAW.8 was much refined in several aspects. In addition to a yaw damper, there was also a pitch damper that had been developed by Glosters and a Sperry autopilot. Aerodynamic changes included alterations to the leading edges of the outer wing sections which were made thinner and exhibited more droop and additional rows of vortex generators on the outer wings to improve the aircraft's buffet characteristics. Developments first seen on the FAW.7 such as the rain dispersal system for the windscreen and the AVPIN engine start were retained. A total of only forty-seven Javelin FAW.8s were built (all by Glosters at Hucclecote) and the first deliveries were made to 41 Squadron at Wattisham. The FAW.8 was also flown by 85 Squadron which was based at West Malling.

Once the Javelin FAW.8 had been tested it became clear that its overall performance and manoeuvrability at high altitudes was significantly better than the un-reheated missile-armed FAW.7. The Javelin FAW.8 was capable of reaching 40,000 ft in four minutes and 50,000 ft in a little over nine minutes. At altitude it could also utilise the increased thrust in reheat to improve turn performance which made the pilot's job of manoeuvring his aircraft onto a target that much easier. Due to the clear benefits that were obtained by the use of reheat the decision was taken to modify a large number of FAW.7s to bring them up to FAW.8 standard and as such they were re-designated as FAW.9s. In all, seventy-six aircraft were subsequently converted to become FAW.9s and the first (XH959) emerged in its new guise on 15 January 1960 [a further twenty-three were FAW.7s that were completed as FAW.9s during manufacture].

The need to deploy fighter aircraft rapidly to bases in the Middle and Far East became a priority in the early 1960s and the Javelin FAW.9 pioneered the use of in-flight refuelling. The earliest trials had been undertaken by an FAW.4 (XA634) which was fitted with a wing-mounted probe, but pilots had experienced difficulty during dry 'hook-ups' and a position was ultimately adopted for a large fixed probe on the starboard side of the forward fuselage. The probe, which measured 20 ft in length, could not be fitted for combat operations and was only used for ferry purposes. To extend the ferry range of the Javelin even further forty-four FAW.9s were modified so that they could carry four 230-gallon fuel tanks attached to the missile pylons under the wings. These aircraft were designated FAW.9R (R for Range). These additional tanks however proved to be a mixed blessing. Several years before some wag had coined the name 'Dragmaster' for the Javelin and the underwing tanks did nothing to improve its smooth passage through the air. Indeed it was just the opposite and there were occasions during in-flight refuelling operations when Javelin pilots had to select reheat just to maintain their position in relation to the tanker. Cynics were quick to point out that the amount of fuel going in was equal to the amount of fuel that was being burnt! In view of this aircraft were often flown with just two tanks instead of four which reduced drag to a slightly more acceptable level.

The Javelin in its developed form was a much more formidable aircraft than the machine that had entered service with 46 Squadron in 1956. For a large aircraft it possessed a superb level of performance (it was superior in many respects to a Hunter F.6) and could climb quickly up to its operational altitude which could be as high as 50,000 ft. It was also an excellent weapons system and its armament of four Firestreak air-to-air missiles could take out hostile aircraft flying at up to 55,000 ft. It was ironic therefore that it started to be phased out of service as an interceptor in the UK in favour of the English Electric Lightning when it had fully matured and had become a very effective all-weather fighter. However it also proved to be extremely adaptable and its large wing area, two-man crew and excellent low-speed handling characteristics made it ideally suited to operations at low level in the Far East. The fact that it was twin-engined was something of a bonus as in addition to the safety aspect when patrolling over inhospitable country, one engine could be shut down if necessary to conserve fuel. Given the aircraft's ability for rapid deployment to anywhere in the world, its imposing presence and weapons capability, the ultimate versions of the Javelin were seen far and wide and although they never became involved in actual conflict, their deterrent effect was of inestimable value.

CHAPTER 12

Javelin Pilot

Having previously flown Mosquito and Meteor night-fighters, Wing Commander Mike 'Dusty' Miller CBE AFC had a long association with the Javelin which included two tours with 23 and 60 Squadrons. As a flight commander with 23 Squadron he was among the first RAF fighter pilots to be involved with in-flight refuelling and later as OC 60 Squadron he oversaw the last operational use of the Javelin in the Far East. Here he gives a personal account of an aircraft he came to know extremely well over a ten year period

> The Javelin was a very easy aircraft to fly as long as you understood its control system and the limitations that were imposed in the looping plane. It handled very well and it was good at height which was mainly down to the delta configuration. For its time it had a lot of power, especially the later marks, and it was reasonably manoeuvrable. It could hold its own against most contemporary aeroplanes, perhaps not quite a Hunter, but it wasn't that far short. At low speed it was really very docile as long as you remembered that you must not stall it and it was adequately provided with stall and

low speed warners to prevent you from doing this. It was an excellent instrument platform and bad weather approaches, even asymmetric, were no trouble at all.

Looping was prohibited according to the book but as long as you knew why this was so there was actually no reason why the aircraft shouldn't be looped. Because of the characteristics of the aeroplane in the stall and the absolute need to avoid a deep stall, the feel simulation aspect of the fully powered controls on the elevators was designed to sense low speed. In simple terms it balanced airspeed pitot pressure against hydraulic pressure and spring pressure in a balancing frame. It was designed so that as the airspeed got dangerously low, the pitot pressure would decay and the springs would take over and push the stick forward. This was fine when the aircraft was the right way up, but if it was at the top of a loop with speed decaying the last thing you wanted was the stick pushed forward on you. This was why looping manoeuvres were banned as it was intended that pilots would not be fooled by the false forward stick pressure.

Provided you understood the system however and knew why it did it and expected it, a loop was a perfectly practical manoeuvre, all you did was to watch your airspeed extremely carefully and make sure that you didn't get too slow. When you were upside down at the top of the loop despite the forward movement of the stick you just kept pulling it back harder to keep the aircraft coming round the loop. There was certainly plenty of power but it took a lot of sky, roughly 10,000 ft, to do a decent loop because you needed to play safe and have plenty of speed at the entry point so that you had enough speed over the top. Obviously you didn't want to try a loop too low and one normally did it around the 20,000 ft level. Having said all that it was illegal to loop a Javelin so of course we didn't do it! Spinning however was completely forbidden and no service pilot as far as

I know deliberately spun a Javelin so the only spinning [other than that carried out by the manufacturer] was done at Boscombe Down.

The Javelin was a supersonic aeroplane, but only just. From the company's point of view it was always boosted as being supersonic but I think the only comment to that is the one by an American who took one look at the Javelin's thick wing and said 'That thing is supersonic? Gee it sure says a lot for thrust!' The best way to get a Javelin supersonic was to roll it on its back and pull as that way it would accelerate quicker. An alternative method was to use full throttle and just push and it would eventually go supersonic in about a 45 degree dive. There were no adverse control aspects at all, all that happened was that the Mach meter tended to hang at just around 1.0 before suddenly jumping to 1.01 and then you knew you were through.

If necessary it was standard operating procedure on Javelins to fly with one engine throttled back to conserve fuel but if you were in dire straits there was no problem in shutting an engine down. The relight system was excellent and I never had any problem relighting. The Javelin was also a good gun platform. The guns were well out on the wings so there was a harmonisation issue which any fighter with wing-mounted guns had. The Firestreak missile was, for its day, an excellent missile. It had a good performance and would give you an increment of 1.7M over your own speed as well as taking out (in theory) up to 10,000 ft in height. Certainly you could expect it to take out 5,000 ft so when you were confronted with a light Vulcan up at 55,000 ft or even more, you could still hack it with a Firestreak-armed Javelin. You had to come in from behind and get within about a 30 degree cone and there were acquisition lights by the gunsight, one for each missile. The Firestreak required quite a demand of services from the parent aircraft with specially-monitored electrics and ammonia to cool the infrared eye.

The first thing you had to do was to arm the missile and it would take two minutes to get the eye cooled. Each missile in its shoe had an ammonia bottle, but every time you switched off and switched on again it took another two minutes to cool the infrared eye and you were running down the ammonia bottle which only lasted fifteen minutes. Generally speaking the missile firing range was three miles down to one mile so within that area and a 30 degree cone you could expect to acquire. Given that you could get the aeroplane into that state, even if it was pointing upwards and losing speed it didn't really matter as long as you had minimum launch speed which was 0.70M. The trick was to get the acquisition, once you had it you could be fairly confident that the missile would take care of your target once it had been launched. As soon as you had launched you could turn away which saved having to get too close to nasty Russian Bears with their 20mm radar-laid guns.

The most difficult target we flew against was the Vulcan, the Victor wasn't quite so difficult and the Valiant was that much easier again. With its pure delta wing the Vulcan couldn't outturn a 4-gun Javelin 7 but it could, if it was light with no bomb load and low on fuel, outturn a Mark 9. This could result in an embarrassing situation for us when 617 Squadron came out to Singapore on exercise. During practice interceptions the CO would take great delight in finishing up on the tail of a Javelin! I don't think he could have got away from a missile attack but if you then went in for a practice gun attack you had to be careful otherwise he could outturn you. Of course this would not have occurred with a fully operational aeroplane with normal bomb load and fuel on board. The Vulcan and the Javelin were, after all, like types as they were both deltas and were in their element at high altitude. The Vulcan had a similar limiting Mach number to the Javelin and I wouldn't mind betting that a Vulcan has been supersonic in its time going downhill.

From the point of view of manoeuvring the best Javelin of all was the 4-gun Mark 7. The reheat on the Mark 9 was a poor compensation given that you had four missiles under the wings although without missiles there was not a lot in it. Above 50,000 ft in a Mark 9 you had to be careful turning above 45 degrees of bank and pulling more than about 2g as you would be into the buffet. Once you got used to the aeroplane it was perfectly reasonable to fly within the buffet as it didn't go immediately into a stall and was still fairly docile. There was actually quite a margin to play with but of course you were losing speed and the book said that when the stall warners sounded you had to recover.

The UK role of the Javelin was essentially as a bomber destroyer aiming to kill Russian bombers over the North Sea. Part of the sophistication of the AI.17 radar was a collimator and once the navigator had locked the radar onto the target, this was then represented by a little green dot on the pilot's gunsight. It only had about a five-degree cone, but it would stay locked out to maybe 25–30 degrees. Until you got it back to within five degrees you weren't sure you were following it exactly, but at least you knew which way to go. With the old AI.10 on the night-fighter Meteor the pilot was flying entirely on instructions from the navigator until he got a visual which could be as close as 300 ft on a really dark night. The biggest problem was approaching too fast and overshooting. On the Javelin, AI.17 made the job of the pilot and navigator a lot easier and the aircraft's powerful airbrakes were a big safety factor. If you closed the throttles and put out the airbrakes in a Javelin you would get a 1g deceleration, as much as jamming the brakes on in a car and locking up the wheels.

The reheat on the Javelin was a very simple system and the Sapphire Sa.7 had minimal modification to provide it. It was designed and intended to compensate for the extra weight of the missile system to try and give the aircraft back the performance of the 4-gun Javelin 7. The guidance was that reheat should not be used below 20,000 ft as below that height, and especially at low level, you actually lost power in reheat. This was because the system was such that the fuel diverted to the reheat actually robbed the main engine of fuel and you lost about 2–3 per cent engine rpm. At height where there was excess fuel pump capacity and fuel was therefore available, the reheat worked in the proper sense. If you had a Vulcan turning on you up at 50,000 ft you needed the reheat to stay with it as it did improve turn performance. All in all though it was a rather poor compensation for having hung four 300 lb missiles under the wings and in some ways it was a bit of a joke to the extent that we used to call it 'wee-heat!'

Air-to-air refuelling on the Javelin was quite a tricky manoeuvre. The drogue of the trailing hose from the tanker had automatic jaws in it which gripped the nozzle of the probe and you had to hit it at less than 7 knots but more than 4 knots overtake speed. This left only 3 knots to play with between making too hard a contact which could cause the hose to 'whip' and too soft a contact in which the jaws in the drogue would not open to allow contact. You approached in a slight climb in line with the trail of the hose but in turbulence the hose could bounce about and make it impossible. In such a situation you usually had to get the tanker to change height and try to find a less turbulent area. This could be difficult, especially at night, because the tanker might not be able to see the best way to go. There were occasions when you would have to hold off and try again a bit later which was always worrying if you were getting short on fuel.

Centre-line closure was the Javelin's Achilles heel in a way. A lot of aircraft (and crews) were lost in the early days over the North Sea and all that was known was that the aircraft had blown up. Quite often the crew didn't have time to get a message out and so aircraft were lost without knowledge for some time. It turned out that in cumulo-nimbus clouds the engines could ingest large quantities of super-cooled water droplets and the engine casing would shrink and touch the compressor blades. Once this happened they broke up and passed back down through the engine wrecking both the compressor and the turbine and bits would fly off quite spectacularly. This was where the Javelin was at a disadvantage as it had two engines close together with hydraulic pumps and gearboxes in the middle. I recall one occasion when F/L Colin Holman was climbing out of Orange in France in a Javelin T.3. He had one engine go and was coping with that when the other one went. He finished up baling out into a thunderstorm and his description afterwards was quite amusing as he wasn't sure whether he was going up or down on his parachute [for an account of this accident see page 111].

Eventually a modification was designed which was a typical Armstrong Siddeley bodge. It was almost literally sticking special, very tough carborundum paper round the casing of the compressor and as the blades touched they just rubbed themselves down. The engine kept going and all you had at the end of the day was a slightly less powerful engine. I did actually experience this in the Far East coming back one evening from our base at Butterworth which was 500 miles up from Singapore. I had to cross the Inter-Tropical Front and as it was clear you could easily see the front and that it went right up to about 60,000 ft. As I couldn't get over the top of it we went through at about 53,000 ft and whilst we were going through both engines suddenly hiccupped and coughed (if a jet engine can cough) before picking up and carrying on. From the cockpit there was just a noticeable deceleration, hiccup and then recovery. When we landed and looked at the jet pipes there was white metal in both. It turned out that both engines had suffered centre-line closure but the modification had worked and the aircraft went on flying.

Although I did tours on Javelins with 23 and 60 Squadrons I only ever fired one Firestreak missile and that was right at the end in Singapore. We were allowed to fire one missile each but we didn't have the drone targets that they had back at Aberporth where you fired at Jindiviks or even old Meteors. The way we operated in Singapore was for a Canberra to drop a flare on a pattern. We then acquired the flare with the missile and knocked the flare out. It was a very demanding exercise as the flare was stationary which was actually outside the parameters of the missile. However it still worked which showed that the missile really was quite capable.

To sum up the Javelin, in many ways it was a much maligned aeroplane but it more than did the job it was designed to do. Although it was a big aeroplane it was more agile than it looked. It acquired something of a reputation because of the prohibition of manoeuvres in the looping plane, which was certainly unusual for a fighter, but it was an excellent bomber destroyer which was what it was really meant to be. It also tackled unusual demands in the Far East in difficult conditions and it really did all it was expected to do, and more.

CHAPTER 13

Accidents and Incidents

During the development of the Javelin and its subsequent RAF service many were lost in accidents, although the overall accident rate was no worse than any other fighter aircraft of the period. Many of the cases involving the Javelin were similar to the incidents occurring to other types, however some were peculiar to the Javelin itself. This chapter looks at a number of the accidents and incidents that occurred, beginning with one of the first examples of an aerodynamic problem that affected a number of aircraft with T-tails

The Deep (or Super) Stall

On 11 June 1953 the second prototype Javelin (WD808) was to be flown by Gloster's deputy Chief Test Pilot Peter Lawrence to obtain a second opinion on the handling qualities of the aircraft after it had been fitted with the modified wing with increased tip chord. Lawrence had only joined Gloster in December 1952 having previously been Chief Test Pilot at Blackburn Aircraft Ltd and had also flown Canberras as part of the test staff at English Electric. He took off from Moreton Valence at 1236 hrs to note the Javelin's characteristics at the approach to the stall, both in the 'clean' and the 'all-down' condition at 20,000 ft. He landed again at 1332 hrs and his aircraft was refuelled and serviced prior to his next sortie. Lawrence took off again at 1526 hrs to continue the stall trials but his aircraft crashed around half an hour later a few yards south of the Old Bristolians cricket field at Long Ashton, Bristol. Although Lawrence attempted to eject he did so at too low an altitude and he was killed instantly when his seat impacted the ground a few yards behind the aircraft.

From an inspection of the crash site it immediately became apparent that the Javelin had hit the ground with virtually no forward speed. This was backed up by a large number of eye witnesses, most of whom reported the aircraft to have been in a tight left-hand spin with very little forward motion. After a brief hesitation during its descent, the left wing was seen to drop again and the aircraft then continued to spin at a high rate of descent until it hit the ground. The weather at the time was fair and played no part in the accident.

As the Javelin was fitted with an automatic observer some data from its final descent was available. The first frame showed the aircraft to be in the normal condition at 125 knots at 15,470 ft and the second frame indicated that it was approaching the stall with flaps up at 103 knots and 13,200 ft. Less than twenty seconds later the Javelin was approaching a 'flaps down' stall at about 95 knots at 13,200 ft and the record then showed the elevators being moved rapidly to the fully down position (24 degrees in 2 seconds) where they remained until just before the crash. During practically the whole descent the aircraft rocked from side to side and also yawed and pitched in a slow, regular manner. At 8,400 ft the rudder and ailerons were held over for seven seconds but this had no effect on the rate of roll in the opposite sense. Engine rpm was fairly low for most of the descent except for a burst of full throttle at 5,400 ft which had the effect of pulling the nose up strongly. The flaps were down during the whole of the descent and airbrakes were not used. Stick and rudder forces were very low to moderate at all stages. The rate of descent was in the order of 8,000 ft/min and the airspeed indicator dropped back to the instrument stop (40 knots or less) at 12,000 ft and remained in this position throughout. The amount of 'g' was low at all times and the aircraft descended at a high angle of attack.

Most of the aircraft was destroyed in the post-crash fire which left only the tailplane section protruding above the charred remains. Even at an early stage in the accident investigation it became apparent that the crash had been caused by some new form of aerodynamic problem. Indeed it was, as this was an example of a 'super stall' in which the high set tailplane was blanketed by the wing during flight at high angles of attack (a situation that was exacerbated by the use of flaps) and rendered the elevators ineffective, thus preventing recovery. This characteristic was to affect a number of other T-tail aircraft in the 1950s and 1960s particularly the BAC One-Eleven and Hawker Siddeley Trident airliners. As far as the Javelin was concerned the 'super stall' was never fully addressed although slotted flaps were developed which reduced the nose-up trim change when these surfaces were lowered. An audio two-tone stall-warning system was also introduced and looping manoeuvres were prohibited.

Centre-line Closure
Throughout the life of the Javelin one of the most serious problems encountered was a critical situation affecting the Sapphire engine known as 'centre-line closure'. A number of aircraft were lost in the early days of Javelin operations and it took some time for the exact cause to be established as vital evidence was either lost or destroyed in post-crash fires. It transpired that in situations when the aircraft was flying through thick cloud there was a distinct possibility that super-cooled water droplets would be ingested into the engines in sufficient quantity as to cause the engine casing to shrink and touch the compressor blades. These then broke

up in a violent explosion which wrecked not just the engine, but hydraulic systems that were routed nearby.

On 9 July 1959 Javelin FAW.7 XH750 of 33 Squadron took off from Wattisham for a transit flight to Middleton St George. A few minutes after take off it flew into a dense cumulonimbus cloud at 8,000 ft and not long afterwards the pilot saw a blue flash in front of him and the aircraft was struck by lightning. A few seconds later the crew felt a violent thump and explosion, and the aircraft went out of control into a spin. The hydraulic warning lights for Nos. 2 and 3 pumps came on and as he looked back the pilot could see that the aircraft was on fire. As the R/T appeared to be dead he used the emergency call light to warn the navigator to abandon the aircraft and both crew ejected safely. The Javelin continued to spin and hit the ground about three miles west of RAF Horsham St Faith.

Although much of the wreckage had been extensively damaged by fire it was apparent that the aircraft had hit the ground after a flat right-hand spin and that a severe fire had existed, particularly in the starboard nacelle, prior to impact. Evidence of lightning strikes were found but these were minor in nature and well away from fuel and oil systems. As a result it was not considered that lightning had any bearing on the in-flight fire. Attention was then switched to the engines, in particular the starboard engine which had stopped when the aircraft hit the ground. There was considerable disruption between the fourth and sixth stages of the compressor and blades and stators had been ejected through the engine casing, some of which were found a mile away from the crash scene. In turn, fuel and oil pipes and installations in the engine bay had suffered serious damage and the hydraulic oil pipes to the tail power controls had also been penetrated by flying debris.

Although the cause of the in-flight explosion was put down to fouling of the fifth stage compressor due to a reduction in clearance, there was a certain amount of disagreement as to why it had happened. Glosters were of the opinion that the weather conditions at the time, i.e. turbulence accompanied with heavy rain (and possibly hail stones) may have had some direct bearing on the failure of the starboard engine. The Accident Investigation Branch (AIB) disagreed with this however stating that there was no direct evidence to support this theory and highlighting the fact that another Javelin which was flying with XH750 at the time had returned safely. The AIB also referred to two other accidents where a Sapphire compressor had failed but where adverse weather conditions appeared not to have been present. In the event the manufacturer was proved to be correct and a modification programme was eventually instigated in which the compressor casing was lined with an abrasive material so that when the blades touched the outer casing they were gradually worn down and disaster was averted.

The identification of the problem and the embodiment of the necessary modification work took a considerable amount of time however and during this period other aircraft were lost including Javelin FAW.9 XH758 of 5 Squadron on 17 October 1963. It was being flown by F/L R. Boulton and F/L L.P. Morley as the leader of a pair, but after take off from Geilenkirchen the aircraft entered cloud at 2,000 ft and continued to fly in cloud and rain up to 27,000 ft. At this altitude there was an explosion and debris and flames could be seen coming from the upper surface of the starboard fuselage in the region of the AVPIN tank. The pilot immediately cut the starboard engine and operated the fire extinguisher before turning towards base. As he did so the controls became ineffective and the nose of the aircraft pitched down into a bunt manoeuvre. After making a short distress call, both crew ejected and their Javelin crashed in flames on a military tank range near Zonhoven. The subsequent investigation revealed that the starboard engine had failed as a result of centre-line closure [F/L Boulton was uninjured but F/L Morley suffered spinal injuries on ejecting].

Less than three weeks later another Javelin crashed due to centre-line closure. On 5 November 1963 a section of two Javelins was transiting through France on its way back to the UK with F/L Colin Holman leading in T.3 XM336. The pair had taken off from Orange and had been cleared to fly not above 4,000 ft for ten miles before climbing to the edge of the control zone under radar surveillance. On leaving the zone the Javelins continued to climb passing though multi-layered cloud, but above 15,000 ft the cloud became much thicker with continuous rain. On reaching 24,000 ft F/L Holman experienced double engine failure and immediately turned back towards the airfield however the rate of descent was such that there was no chance of a successful recovery and both crew ejected at a height of 6,000 ft. F/L Holman sustained a back injury and his navigator (F/L D.E. Berks) bruised a foot on landing. Tragically the Javelin came down on a row of houses near Saujac and four people were killed, with nine injured.

Other Compressor Failures
Problems were also experienced with the Sapphire compressor when centre-line closure was not the cause. On 11 September 1963 Javelin FAW.8 XJ113 of 41 Squadron was about to take off from Wattisham for a continuation training detail. After lining up with the runway the pilot opened up both engines to 80 per cent rpm against the brakes before checking that all temperatures and pressures were normal. The brakes were then released and power opened up to 100 per cent rpm with the JPT gauges showing a healthy 655 degrees C. As the speed reached 30–40 knots however there was a loud bang and the aircraft juddered violently. The pilot closed both throttles and began braking but the aircraft then swung violently to port and despite the application of full starboard brake it could not be prevented from leaving the runway. It eventually came to rest on the grass about fifteen yards from the edge of the runway.

The pilot immediately suspected that a tyre had burst and he informed the tower of what he thought had happened. He then closed both HP cocks and switched off all the electrics before instructing the navigator to leave the aircraft. As the navigator looked back however he saw smoke and advised the pilot to evacuate as quickly as possible. Once on the ground the crew members saw that the starboard ventral tank was on fire and that there was smoke billowing from the starboard engine intake and both jet pipes. The fire crew soon arrived and managed to contain the fire around the ventral tank but as soon as they moved away to try to extinguish the fires elsewhere, the ventral tank erupted once again. It also became apparent that a severe fire was raging in the starboard engine and this quickly spread to other areas. By this time the ammunition and ejector seat cartridges had begun to explode so that it was too dangerous for the fire crews to remain on station. Having retired to a safe distance the Javelin had to be left to burn itself out. When the wreck was eventually examined it was revealed that the accident was most probably caused by failure of the first stage compressor blades on the starboard engine. The violent disintegration of the blades had severed fuel and hydraulic lines which led to the uncontrolled fire.

An almost identical accident occurred on 30 January 1964 and involved Javelin FAW.9 XH723 of 29 Squadron during a deployment to Nicosia in Cyprus. The pilot had been briefed to fly as No. 2 to another Javelin and both aircraft lined up on the runway for a stream take off. After opening up to 80 per cent power for control checks, the pilot released the brakes before advancing the throttles to full power. He had only moved a few yards forward when he heard a muffled explosion and the aircraft began to vibrate. His immediate actions were to close the HP cocks, at the same time as bringing the aircraft to a halt. He too suspected that a tyre had burst and informed the tower of this, however the controller radioed back to say that the Javelin was on fire [there had been no fire warnings in the cockpit]. The crew quickly left the aircraft but the fire developed rapidly and appeared to be centred in the starboard engine bay. The starboard ventral tank was also on fire but by this time it had become detached and was lying on the runway. Due to the severity of the conflagration the Javelin was damaged beyond repair before fire crews could bring the flames under control. The cause of the accident was much the same as the accident at Wattisham as the 1st stage compressor of the starboard engine had failed and this in turn had ruptured fuel and hydraulic pipes leading to a fire that was aggravated by fuel from the starboard ventral tank.

Compressor problems were also experienced in the Far East towards the end of the Javelin's service life when the aircraft spent much of its time at low level. The cause was eventually put down to compressor blade vibration when the engines were operated in a particular rpm range. Several aircraft were lost including FAW.9 XH877 of 64 Squadron on 22 June 1965. It was being flown by F/L P.J. Hart on a patrol at 1,000 ft but when he opened the throttles after forty-five minutes in the air there was a muffled explosion from the port engine. At the same time the aircraft began rolling to port in a nose-up attitude and did not respond to corrective action. It also began to vibrate violently and as it became inverted the crew ejected. The cause of the accident was failure of the port engine compressor and the remains of a number of blades were found some way from the main crash site.

In-Flight Fires

In addition to centre-line closure and failures of the compressor, problems were also experienced with the turbine section of the Sapphire engine and a number of in-flight fires were the result. On 11 February 1958 F/O F.H.B. Stark of 23 Squadron (with F/O P. Baigent as navigator) took off from Coltishall in Javelin FAW.4 XA734. Shortly after becoming airborne a severe fire broke out in the rear fuselage and there was a loss of hydraulic power. The crew abandoned the aircraft successfully and were uninjured. The Javelin came down sixteen miles south-west of Coltishall, virtually in line with the take off runway, however the fin and tailplane had broken away and were found 600 yards from the main crash site.

The investigation into this accident centred on the starboard engine when it was noted that the first stage turbine blades were missing and the first and second stage stator rings had burst open. The accident was eventually put down to a fatigue failure of the starboard engine outer flame tube casing which had caused significant overheating of the first stage turbine blades. Flying debris from the turbine blades had, in turn, ruptured fuel and hydraulic systems and caused a fire severe enough to burn away the fin and tailplane attachment. Modifications were already in hand to further improve and strengthen the turbine of the Sapphire (Mods.389 and 390) but neither had been embodied on XA734 at the time. Following this accident the Sapphire engine was temporarily limited to 8,200 rpm with a Jet Pipe Temperature (JPT) of 595 degrees C.

A similar accident took place on 29 September 1959 and involved Javelin FAW.5 XA662 of 228 OCU. The aircraft was being flown by F/O C.P. Cowper with Capt R. Neitz (USAF) in the rear cockpit. After take off from Leeming the Javelin was climbed to the west and was passing 32,000 ft when F/O Cowper noticed that the starboard engine was starting to run down. Assuming flame extinction he pressed the relight button but this had no effect and the engine continued to spool down. A vibration then developed which became progressively worse and a fire warning illuminated. The light went out when the fire extinguisher was operated but as Cowper was turning the aircraft to return to base the port engine fire warning came on. This time the fire extinguisher had no effect and the warning light stayed on. He ordered the navigator to eject and then abandoned the aircraft himself. Both were uninjured. The Javelin crashed on the moors near

Leyburn in Yorkshire and set fire to peat which burned for several miles.

Although the aircraft had disintegrated on impact with the ground due to the explosion of the large quantity of fuel on board, enough wreckage was recovered to give an idea of what had happened. Fire damage was severe on the starboard side and it appeared as though there had been a failure of the primary turbine in the starboard engine. A closer examination showed that a number of blades in this area had failed in very rapid creep and exhibited abnormal hot tensile fractures, indicating that at some time during its life, maximum design temperature had been exceeded. Although the life of Sapphire engines between re-conditioning was 400 hours, blade life was 800 hours, and because of this accident it was recommended that, in future, blade life should be the same as engine life.

Despite modifications to the Sapphire turbine, accidents continued to happen. On 7 June 1962 Javelin FAW.5 XA645 took off from Laarbruch at 2221 hrs on a battle flight scramble. It was being flown by F/L J.H. Adam with F/L C.M. Pinker in the rear seat. When at a height of 2,000 ft F/L Adam noticed that both JPT gauges were reading 660 degrees C and throttled back to 97 per cent which gave JPT readings of 640 degrees C. The climb was continued but at 9,000 ft there was a muffled explosion which shook the entire aircraft. Almost immediately both engine fire warning lights came on. After throttling back the engines the fire warning lights went out but this was accompanied by a progressive dimming of the cockpit lights. By now flames could be seen coming from the rear of the aircraft and the fire continued after the fire extinguisher bottles were operated. The navigator was ordered to eject at 9,000 ft and the pilot followed. Both escaped unhurt and the Javelin crashed in a field approximately twelve miles west of Dingden.

The engines were recovered from the crash site but in this case the damage sustained by the starboard engine was consistent with ground impact at low idling rpm. The port engine however revealed evidence of a mechanical failure in the air which had, once again, damaged fuel lines and led to an uncontrollable fire. The second stage turbine blades had sheared and this was considered to be the primary failure (the first stage turbine blades had also detached but this was thought to be a secondary effect). The reason for this failure was put down to overheating due to a reduction in cooling air caused by 'bottoming' of the combustion chamber inner flame tube rear seal. It was calculated that the temperature in this area had been in excess of 725 degrees C. During the crash investigation the engine manufacturers stated that around 40 per cent of engines returned for repair exhibited signs of overheating in the turbine area. Eventually a modification was embodied (Mod.810) but this was not a complete cure and restrictions leading to overheating still occurred.

Inadvertent Ejections

There were several cases of pilots being inadvertently ejected from Javelins and this occurred for the first time on 26 February 1958 when FAW.7 XH714 was being flown by F/L R.S. May from 'A' Squadron of A&AEE Boscombe Down on a routine test flight after major servicing. Shortly after crossing the coast near Bournemouth the pilot was ejected from the aircraft at an altitude of about 5,000 ft and a speed of 400 kts IAS. The aircraft went into a spiral dive and the navigator (F/O J.M.V. Coates) ejected at around 2,000 ft. Both he and the pilot however were killed.

It was found that the pilot had been inadvertently ejected from the aircraft because of an incorrectly assembled top lock on his Mark 3JS ejection seat. This meant that the seat was not locked to the aircraft at any time. Under slight negative 'g' the seat rose, firing the drogue gun through the canopy and extracting the drogue, thus initiating the sequence that led to the pilot being ejected through the canopy. The seat trajectory was abnormal and the seat itself was damaged, possibly as a result of striking the tail and fin, so that the parachute only partially deployed. As the navigator had not immediately ejected after the pilot, the AIB expressed the opinion that severe buffeting in the rear cockpit may have delayed his departure. The Javelin's speed and rate of descent when he ejected were unknown. Separation from the seat took place but there was insufficient time or altitude for the auxiliary parachute to deploy the main canopy. There was evidence that the navigator's seat had fouled part of the cockpit structure during ejection so that his parachute would have taken longer to deploy.

It appeared that the damage to the starboard side of the pilot's seat had prevented the starboard lap strap lock from releasing as the seat remained attached at this point and had flailed around him (the port lap strap and shoulder harness lock had released satisfactorily). Although the parachute did deploy, it 'candled' as the parachute apex remained attached to the seat. The Javelin crashed at Sandford near Ringwood in Hampshire. The pilot was located with his ejector seat seven miles from the crash site and the navigator's body was found ¼ mile from the aircraft crater.

On 18 February 1959 F/L R.A.J 'Dick' Carrey of 87 Squadron was flying Javelin FAW.1 XA569 on a high-altitude training flight from RAF Bruggen when he was ejected through the canopy. Once again this was caused by the top lock of the ejection seat being incorrectly assembled. Dick Carrey later described the aftermath of his ejection in a letter to Flight International magazine

I had a very rough ride in the ejection seat down to the auto-separation altitude. The drogue assembly had been damaged in exiting through the plexiglas canopy, with the result that the seat was spinning and somersaulting all the way down from 42,000 ft. My relief when the automatics operated perfectly turned to consternation when I pitched forward, not only out of the seat, but out of my parachute harness as well. Luckily for me one of the thigh straps jammed in the crotch loop and I ended up at 10,000 ft hanging upside down by one leg. I must confess that this was a bad moment. By hauling myself upright by the dinghy pack I was able to hook my arm through the harness and clasp my hands together before the jammed thigh strap gave way. By the time I landed on a small island in the Scheldt estuary, I seemed to have been hanging from that arm for a very long time.

Sadly Dick Carrey's navigator, F/L Alec Cooper, was killed. The drogue gun bullet of his ejection seat was either not connected to the drogue withdrawal line or became detached at some point during the ejection sequence and Cooper's parachute did not open.

Mid-Air Break Up

During its service life the Javelin was subject to a normal acceleration limit of 5.5g and to try to ensure that this figure was not exceeded the stick force per 'g' was approximately 7 lbs, in theory making it difficult for a pilot to haul back on the control column excessively. However if an aircraft was being flown out of trim it was quite possible to bring about high normal accelerations without undue force being applied to the stick. On 29 August 1961 Javelin FAW.8 XH971 of 41 Squadron was being flown by F/L J.L. Hatch, with F/L J.C.P. Northall as navigator, and was returning to Geilenkirchen. The Javelin approached the airfield at a speed of 460 knots at a height of 400 ft and carried out a break manoeuvre at the upwind end of the runway, almost directly opposite the squadron dispersal. As it did so the aircraft broke up.

There were many eye witnesses but the disintegration happened so quickly that the sequence was difficult to piece together (a cloud of vaporising fuel was also produced). It was thought that the port wing had failed first and that this had been followed by the cockpit and front fuselage separating from the remainder of the aircraft. At some time the starboard and port wings clapped together. The ejection seats separated from the Javelin but both pilot and navigator were killed.

As part of the crash investigation, trials were conducted to establish the longitudinal stick force per 'g' and it was found that the force became increasingly light as acceleration increased. It was calculated that the difference in the stick force between the service limit and the amount of 'g' pulled in the accident (approx 10–11g) was only about 10 lbs. This was only part of the story however as it transpired that at the time of the accident XH971 was being flown in a trim state corresponding to a speed of 300–330 knots. As the actual speed was 460 knots the pilot would have needed to apply a push force on the stick of around 10 lbs to maintain level flight. If the Javelin had been flying in trim at 300 knots a pull force of 30 lbs would have only produced about 5.5g however the 'g' loading generated by converting a 10 lb push to a 30 lb pull was sufficient to cause a structural failure [a rather different type of structural failure occurred on 10 February 1964 and involved Javelin FAW.9 XH747 of 60 Squadron which was based at Tengah in Singapore. On this occasion the entire tail section broke away and the aircraft crashed in the sea with the crew ejecting. For an account of this accident see page 178]

Electrical Failure

On 19 August 1959 F/L A.D. Lewis of 41 Squadron was to have carried out a practice sortie in Javelin FAW.5 XA667. He was soon into an emergency situation however and the following rather understated account produced for the Court of Enquiry belies an excellent piece of flying

I was detailed for a high level practice interception and took off from RAF Wattisham at 2035 hrs. Owing to brake trouble on the second aircraft, I took off alone and climbed to 8,000 ft and orbited the airfield to port. After about 3–4 minutes I felt a sudden thud which shook the aeroplane and was similar to crossing another aircraft's slipstream. Immediately following the thump, I noticed that the port generator warning light was on (this is a red light situated on the right-hand side of the front instrument panel). I told the navigator to switch off all his equipment including the AI and IFF. This left a load on of about 20 amps to run the Gee which, at this time, was still running satisfactorily.

About four minutes after the port red light had come on, it went out and a total electrical failure occurred. During that time I had been flying with fully extended airbrakes and using about 90 per cent power on the engines in order to burn up fuel to get down to landing weight. When the electrical failure occurred I lost all cockpit lighting so switched on the emergency lighting for landing. As the gyros slowed down I lost the navigation instruments, compass, artificial horizon etc and I was unable to use tail trim. In order to reduce weight as soon as possible I decided to jettison the two ventral tanks. I flew out into the Harwich area reducing height to 4,000 ft and speed to approximately 270 knots and when four miles off the coast and heading in an easterly direction I jettisoned both tanks. As the tanks left the nose of the aircraft rose slightly.

When I finally came into land after having dropped the ventrals I had no radio communications with the ground so I deliberately desynchronised the engines to draw attention to the fact that I was in the circuit. I landed the aircraft at 2105 hrs having been airborne for about thirty minutes. I had no undercarriage indicator lights to tell me whether the undercarriage was fully locked down for landing. I came over the airfield boundary at 140 knots to allow for the excess weight and landed normally.

After the Javelin landed it was immediately apparent that a servicing door was hanging loose and was badly damaged.

Buffeting caused by the airflow had tripped the inertia switches for the Graviner fire extinguisher system, activating the fire extinguisher bottles and this had led to electrical isolation. The incident was caused by the servicing panel not being securely fastened when the aircraft took off so that it tore open in flight.

Landing Accidents

In the early days of Javelin operations a serious design weakness was that rain on the windscreen virtually obliterated all forward vision, a situation that was particularly hazardous when landing at night. A number of minor landing accidents occurred which also highlighted a characteristic of the Javelin in that at low speeds the elevators were not very effective and so a bounce or stall near the ground was likely to cause the aircraft to porpoise. When the engines were at idling it took some time before there was any appreciable increase in power after opening the throttles. Because of this it was difficult to synchronise the power selection with the attitude of the aircraft and the porpoising could be accentuated. If porpoising did occur it was best not to open the throttles and let the effect dampen out as speed was lost.

Brake operation also caused problems at first as many pilots had no experience of toe-operated brakes. The brake system on the Javelin also left a lot to be desired as the slightest unevenness of braking in either main wheel caused directional instability. Experience soon showed, however, that instead of using full flap in crosswind conditions, it was better to use partial flap and full airbrake which provided the equivalent drag but reduced the tendency to weathercock. This in turn minimised the need for differential braking.

During the service life of the Javelin there were many cases of undercarriage malfunction and a typical example occurred on 22 February 1960 when FAW.4 XA763 of 3 Squadron was about to land at Geilenkirchen. During a GCA the pilot selected undercarriage down when ten miles from the runway but the starboard leg would not lower. Subsequent re-selections and use of the emergency system failed to have any effect and so the pilot had to land on the port and nose wheels. It appeared that the accident had been caused by nothing more than a sealing ring breaking up and entering the hydraulic system, creating a temporary blockage in the pressure line to the starboard leg.

Some aircraft had persistent failures and Javelin T.3 XH443 of the Fighter Command Instrument Rating Squadron was involved in two landing accidents in the space of seven months. The first occurred on 5 September 1962 during a standardisation check and occurred after three successful circuits and landings had been made. When the undercarriage was next selected down the starboard red light stayed on and remained so despite re-selections, manoeuvring involving positive and negative 'g' and use of both pilots' emergency air system. The captain decided to land the aircraft from the rear seat and touch down was made at 125 kts IAS with half flap and airbrakes selected out. The wing was held up for as long as possible and the aircraft was kept straight after it touched by using the rudder and brake. It transpired that the undercarriage had failed to lower due to a faulty release unit. XH443 was in trouble again on 18 March 1963 during a dual instructional detail involving circuits and landings. On this occasion it was the port leg that refused to come down and this was confirmed by a visual check by the controller. Once again all attempts to get the offending leg to lower were unsuccessful and a gentle landing was made on the nose wheel and starboard main wheel. By an excellent piece of flying the damage to the aircraft was relatively minor.

There were also many instances of burst tyres on landing which quite often ended with collapse of the undercarriage. A typical example took place on 20 September 1960 when FAW.7 XH838 of 33 Squadron was landing at Middleton St George. The port tyre burst and the undercarriage leg fractured at the top attachment point, causing it to break away. It was noted at the time that this was the tenth case in recent months that a Javelin tyre had burst on touch down and in all of these incidents no technical fault had been found.

Landing in a crosswind could also cause problems as a student pilot discovered on his first solo flight in a Javelin FAW.5 (XA701) at Leeming on 5 May 1961. As the main runway at Leeming was roughly parallel with the Great North Road (A1) which ran alongside the aerodrome, the prevailing westerly wind often generated a substantial cross-wind component. On this occasion the crosswind was recorded at 22 knots (rather more than when the pilot had taken off) and shortly after touching down at 125 kts IAS the starboard tyre burst and the aircraft started to swing gently to starboard. Despite the application of full port rudder and port brake, the pilot was unable to prevent the aircraft from leaving the runway. Once onto the grass the aircraft ground-looped and both undercarriage legs broke away. It was later concluded that the burst tyre had been caused by incorrect use of the brakes by the pilot. However operation of the braking system on the Javelin had been adversely commented upon in the course of numerous flight trials and needed a careful touch. This was particularly so when landing with a strong crosswind.

Jet-pipe Fractures
During the Javelin's early service there were a number of instances of jet-pipe fracture which was a particular concern as hot gases from the jet efflux came into contact with areas of the centre fuselage where there was no fire warning. This occurred to FAW.5 XA706 of 228 OCU on 29 June 1960, the pilot experiencing progressive failures of the hydraulics, R/T, navigation lights, fuel gauges and fuel flow meters. An immediate return was made to Leeming where waiting fire crews were able to extinguish the fire. A similar incident occurred to FAW.2 XA813 of 46 Squadron on 12 April 1961.

Around an hour after take off the No. 2 hydraulic warning light came on and during the approach to an emergency landing at Waterbeach this was joined for a short time by the No. 3 warning light. The aircraft landed safely where it was discovered that a fractured jet-pipe had allowed hot gases to escape and this had caused buckling of the keel plates. As a result of severe overheating a hydraulic union had failed and this had led to a serious leak of hydraulic fluid which had then vaporised and caused a fire. Other incidents of this nature affected FAW.5 XA701 of the All-Weather Fighter Combat School on 28 November 1961 and T.3 XM336 of the Fighter Command Instrument Rating Squadron on 10 August 1962. In both cases safe landings were made although the aircraft were damaged Cat.3 [by the time of the latter incident a modification had been incorporated which was meant to prevent this type of hot gas leak].

Hydraulic Failures
Although the Javelin did not suffer from hydraulic failures any more than contemporary aircraft, malfunction of the hydraulic systems led to a number of aircraft being written off and the following examples illustrate the type of incident that Javelin crews had to be prepared for. On 8 September 1958 a pair of Javelin FAW.4s of 41 Squadron took off from Wattisham to carry out mutual practice interceptions. Shortly after becoming airborne the pilot of the No. 2 Javelin noticed that the fore-and-aft stick forces were lighter than normal and that he could not trim out the stick forward load by trimming nose-up in the climb. He levelled out at 10,000 ft in clear air to carry out checks during which it became apparent that the tail trim indicator was reading full nose-up. Within a second or so of levelling out the aircraft pitched up violently to an angle of approximately 40 degrees and it flick rolled to port through 360 degrees. By pushing hard on the control column the pilot managed to regain normal flight, whereupon he noticed that the trim indicator was

124

reading one division nose-up trim. After asking his No. 1 to check for signs of damage to his aircraft, he selected nose-down trim on the emergency trim switch and returned to base without further problem. This incident was put down to malfunctioning of the powered controls artificial feel system brought about by contamination of the hydraulic oil system by foreign particles.

On 22 October 1959 a ferry pilot was taking off in Javelin FAW.9 XH761 from Moreton Valence when he noticed that the red warning lights for Nos. 2 and 3 hydraulic systems operating the flying controls were illuminated. At this point he was about half way down the runway at 135 kts IAS and was just about to get airborne. Making an instant decision, he abandoned the take off and closed the throttles, selected airbrakes out and applied maximum wheel braking. As there was only 900 yards of runway remaining when he elected to abort his take off the Javelin was soon into the overshoot area and went through the boundary hedge. It then passed over a road at which point the undercarriage broke off when the wheels hit a small bank on the verge and it slid into an adjacent field, finally coming to rest after a further 100 yards. Both ventral fuel tanks broke away as the undercarriage collapsed but there was no fire and the pilot, who was un-injured, managed to escape by utilising the front canopy jettison system. The aircraft, which was virtually brand new and had only flown around four hours on contractor's test flights, was damaged beyond repair.

The two red warning lights indicated that hydraulic pressure had fallen below 2,500 psi. At the time of the accident if a pilot was confronted with one warning light he was advised to restrict control movements and return to base, but if both red lights illuminated he was to abandon the aircraft. When the hydraulic system was examined only one system was found to be faulty and the reason that the other red light came on as well could not be established. A number of spurious warnings had already occurred in service but not at such a critical moment as on this occasion.

Although it transpired that the pilot could have taken off and retained control, the fact that he was confronted with two red warning lights meant that his only course of action was to abandon the take off.

Another Javelin FAW.9 to be written off as a result of hydraulic failure was XH794 of 33 Squadron which crashed on landing at Wildenrath on 9 March 1962. The aircraft was being air tested by S/L S. Burrows to check the hydraulic power controls following a component change and had just carried out a GCA approach at Wildenrath when the pilot declared an emergency. At this stage the No. 3 hydraulic warning light came on and ground observers saw the aircraft turn to starboard and lose height rapidly. During the last part of this turn S/L Burrows later remembered having great difficulty in moving the control column to prevent the aircraft from striking the ground. The Javelin overshot the runway and hit a telegraph pole with its starboard undercarriage, touching down heavily in a level attitude just short of the airfield boundary fence. It crossed the boundary and finally came to rest upside down about 350 yards from the first point of impact. S/L Burrow's ejection seat fired at some point after the aircraft hit the ground and he was found, still strapped to his seat, about seventy yards from the remains of his aircraft. The navigator was trapped in the wreckage and it took over four hours for him to be freed. Both pilot and navigator survived, albeit seriously injured. It was later concluded that the loss of control was the result of a severe leak affecting the Nos. 2 and 3 hydraulic systems.

Start-up Fires

Starting a Javelin could be a hazardous operation and there were some spectacular conflagrations when things went wrong. The Sapphire engines fitted to the early Javelins were started by a cartridge approximately six inches long which contained cordite as an explosive charge. There was provision for two cartridges to be loaded for each engine (one being a reserve) and they were fired electrically. As the cartridge operated, the cordite gases expanded and were fed to a turbine starter which would spool the engine up to 2,000 rpm before disengaging. That, at least, was what was supposed to happen.

One of the most impressive blow-ups took place at Geilenkirchen on 29 October 1962 and involved Javelin FAW.5 XA661 of 11 Squadron. After firing the first cartridge on start-up, the engine rpm increased to only 12 per cent and the engine failed to light. The ground crew then checked the servicing bay and jet pipe for signs of damage or fire and on finding none they indicated to the pilot that he should attempt a re-start. When the second cartridge was fired however there was an almost instantaneous explosion and flames erupted from above and below the port engine in the vicinity of the starter motor. The pilot and navigator quickly abandoned the aircraft as the ground crew attempted to tackle the fire with a portable fire extinguisher, but this had no effect. They were then forced to retire to a safe distance as there was a danger of the fuel tanks erupting. It was not until the arrival of the fire tender that the fire could be brought under control but by this time it had caused damage sufficient for the aircraft to be written off. The accident was put down to failure of the starter turbine disc which, after disintegrating, penetrated the No. 1 fuselage fuel tank and ignited the fuel.

This was the second Javelin to be damaged Cat.5 during October 1962 as FAW.5 XA701 of the All-Weather Fighter Combat School at West Raynham was damaged beyond repair on the 4th. The sequence of events was very similar as a loud bang was heard shortly after the pilot attempted to start the port engine. On this occasion the ground crew were able to put out the fire which had started in the aircraft's servicing bay but not before substantial damage had been caused. The accident was cause by a blocked pipe preventing the release of excess pressure which then damaged the unselected cartridge sufficiently for it to explode.

On one occasion the engine starting system on the Javelin contrived to go awry after flight. Having carried out a training sortie on 4 May 1961, the crew of FAW.5 XH692 of 228 OCU at Leeming were engaged on their after landing checks when the remaining starter cartridge for the starboard engine inadvertently fired. This caused the starter to disintegrate and fragments pierced fuel tanks in the starboard fuselage and wing. The aircraft caught fire and after a short period the starboard ventral tank exploded. The crew were able to evacuate the aircraft safely but the resultant damage led to yet another write off.

The later marks of Javelin used a different starting system whereby a much smaller cartridge was used to ignite AVPIN which was a highly volatile fuel. The expanding gases from the burning AVPIN then spun the turbine which, as before, activated the engine. The new system was just as hazardous however as illustrated by a start-up fire that occurred on 9 April 1962. Javelin FAW.8 XH977 of 41 Squadron was being started but after the starboard engine starter button was pressed a muffled explosion was heard and the pilot immediately switched off the electrics. It was only the prompt action of the ground crew in extinguishing the fire in the servicing bay that saved the aircraft from being seriously damaged. A fractured pipe had allowed hot gases to ignite electric cables.

Mid-air Collisions

The first mid-air collision involving a Javelin occurred on 24 August 1956 when FAW.4 XA644 which had been retained by Glosters for testing collided with Hunter F.4 XF980 of the Examining Wing at Central Flying School. The pilot of the Hunter was F/L Peter Hicks who later described what happened

On my third sortie [on the Hunter] I took off on the westerly runway at Kemble and climbed straight ahead at 430 knots. I entered cloud momentarily at about 10,000 ft

128

and very soon shot into the clear. Simultaneously there was the mother-and-father of bangs. A 'bang' is a pretty inadequate description of 15,000 lbs of Hunter at 430 knots impacting 38,000 lbs of Javelin at, I believe, 500 knots descending. In the event, the Hunter broke in two somewhere aft of the cockpit, and the navigator of the Javelin who also survived, saw the floor splitting beneath his feet. With the bit of Hunter I was sitting in tumbling and rolling, there was little doubt as to what to do next. I was very lucky to be able to do it. Afterwards it was established that my bone dome was severely smashed in on the left side by a piece of aircraft structure and that there was a hole in the seat frame about six inches behind my ear. Nice timing at the closing speeds involved!

Peter Hicks suffered severe leg injuries as his seat was damaged during the ejection process and the drogue did not operate. As a result his legs flailed badly as his seat tumbled through the air and it would be six months before he could return to CFS. Both of the Javelin seats were fired out in the break up, the canopy explosive jacks and ejection guns operating normally. The navigator of the Javelin who survived was F/L R.E. Jeffries and was on attachment to the manufacturer. He suffered only a minor arm injury, however Gloster test pilot Brian Smith was killed. It appeared that his parachute rigging lines had come into contact with aircraft structure as they had all been cut through. The fact that two of the aircrew involved in this accident survived was remarkable considering the nature of the collision.

Another mid-air collision occurred on 21 May 1960 and involved Javelin FAW.6s XA823 and XA835 of 29 Squadron. After a mass scramble from their base at Leuchars, both aircraft climbed to 35,000 ft on a heading of 160 degrees which took them over the North Sea. The crew of XA835 (pilot F/L D.J. Wyborn and navigator F/L J.S. Clark) then turned onto a course of 110 degrees and climbed to the holding point at 42,000 ft. Just after levelling out the aircraft was hit from below by XA823 but F/L Wyborn managed to maintain control and flew to the coast where the crew ejected safely. The aircraft crashed in open country to the north-east of Scarborough. XA823 was more seriously damaged and went into an uncontrollable dive to crash in the sea off Hartlepool. The crew (pilot F/L J.S. Wilson and navigator F/O E. Wood) also ejected and were later rescued from the sea.

Practice interceptions at night could be particularly hazardous and on 26 October 1961 Javelin FAW.9 XH906 of 25 Squadron collided with Canberra B.2 WD995 of 32 Squadron during a night low level interception exercise near Akrotiri in Cyprus. The crew of the Javelin were unable to establish AI contact with the target and the pilot decided that he would make a visual attack. At an estimated range of ½ mile the Javelin rolled out of a turn that placed it directly behind the target and then began to descend. At night it could be extremely difficult to gauge closing speeds when making visual attacks and on this occasion the pilot mis-judged how close he was getting to his intended target so that the starboard wing of the Javelin struck the tail fin and port elevator of the Canberra. The two aircraft crashed and the pilot of the Javelin (F/L John Morris) and the three-man crew of the Canberra were killed. The navigator of the Javelin (P/O Roger Lloyd) was able to eject and survived.

Sadly this was not the last mid-air collision involving Javelins as two aircraft from 64 Squadron crashed about ten miles north-west of Tengah on 30 May 1967. The accident occurred during a low level close formation exercise involving six aircraft and took place shortly before 64 Squadron dis-banded. During the join-up at a height of about 1,000 ft, it appeared that the pilot of the aircraft in the No. 3 position selected airbrakes without warning the No. 4 which was directly behind him. This led to the No. 4 overshooting so that it ended up ahead and below the No. 3. In attempting to regain formation the No. 4 collided with the other Javelin and both aircraft went out of control. The crew of the No. 3 (XH896) were able to eject but the No. 4 (XH708) dived straight into the ground where it exploded on impact and both occupants were killed. These were F/O W.D. Kay and a member of 64 Squadron's ground crew, Cpl Kenneth Ashbee.

CHAPTER 14

Javelin Squadron

Number 64 Squadron was formed at Sedgford on 1 August 1916 as a training unit but the following year it moved to France where it flew the DH.5 and SE.5a on the western front. After being disbanded in December 1919, the squadron was reformed on 1 March 1936 with Hawker Demons, these being replaced by Bristol Blenheim If night-fighters in 1939. Shortly before the Battle of Britain 64 Squadron changed roles to become a day fighter squadron and flew various marks of the Supermarine Spitfire until November 1944 when it received North American Mustang IIIs. After the Second World War 64 Squadron flew de Havilland Hornets and Gloster Meteor F.8s but reverted to the night-fighter role in 1957 and flew the Meteor NF.12/14 until August 1958 when it became non-operational for conversion to the Javelin FAW.7. By this time 64 Squadron was based at Duxford and was under the command of W/C J.A. Wright.

The first three weeks of September were occupied by lectures and demonstrations by the Javelin Mobile Training Unit (JMTU) and the remainder of the month was taken up by familiarisation sorties in the Javelin FAW.7, of which six had arrived by the end of the month. Conversion was continued the following month although this process was hampered by poor aircraft serviceability and bad weather. Even so most crews managed to complete their air exercises and were ready to commence night flying. On 11 November the squadron moved to Stradishall to continue working up on the Javelin but due to bad weather, flying had to be restricted to day sorties however a considerable number of Practice Interceptions (PI's) were flown.

On 22 January 1959 the squadron returned to Duxford where the weather proved to be kinder than at Stradishall and the first full night flying programme was completed. Poor aircraft serviceability was still an issue, mainly due to lack of spares, but it was hoped that there would be an improvement during the following month. This turned out to be the case and with improved weather as well, 64 Squadron managed to complete nine full night flying programmes and started cine gun exercises in preparation for air firing. During February the last Javelin FAW.7 was delivered, bringing the squadron up to a full complement of sixteen. March was the busiest month to date with the Javelin with the first half of the month given over to normal training flying by day and night, with every crew having at least one cine flag sortie. On the 16th a two-week period of air firing was commenced after which the squadron was declared fully operational. Total flying for the month amounted to 164.20 hours by day and 79.30 hours by night.

Having become operational once more 64 Squadron took part in its first exercise with Javelins on 2 April (Exercise Buckboard) and eleven scrambles produced a healthy total of twenty-two 'splashes'. On the 13th seven aircraft were provided to simulate bombing attacks on the continent for Exercise Topweight and that night acted as fighters during the night phase. Training was continued during May with PI's and air firing to the fore, the squadron also providing hospitality to members of staff from 'The Aeroplane' magazine who interviewed a number of personnel and took photographs for a forthcoming article. Activity during June was much the same although thoughts were turned increasingly to an upcoming detachment to Nicosia and the last week of the month was mainly taken up with air tests and equipping the aircraft with Marconi Radio Compasses for the long journey. Eleven Javelins departed on 30 June for Luqa in Malta, en-route to Nicosia for Exercise Quickfire and arrived safely on 1 July. The detachment was invaluable for training purposes and the squadron achieved its highest monthly total of flying hours yet with just over 425 hours by day and night. Most of the flying was devoted to ground controlled interceptions and mutual PI's, however the squadron also exercised with Canberras and local naval units. Battle flight was also provided and four scrambles were called.

During the detachment to Cyprus Javelin FAW 7 XH789 'G' was written off after it overran the runway during an emergency landing at Akrotiri on 30 July. It was being flown by F/L Colin Grindley with Sgt Steve Sanders as his navigator but at a height of 35,000 ft a double hydraulic failure was experienced which meant that hydraulic control to the undercarriage, flaps and airbrakes was lost. The only way the undercarriage could be lowered was by the emergency air bottle and the crew elected to do this and try to land the aircraft, rather than eject. With no flaps and airbrakes however the Javelin left the runway still doing around 90 knots and finally came to rest in the surrounding countryside. Although the Javelin was damaged Cat.5, Grindley and Sanders were uninjured and were able to vacate their aircraft safely.

Exercise Quickfire ended on 6 August and the Javelins of 64 Squadron returned to Duxford the following day to be stood down for a week. Training re-commenced on the 17th although weapons training suffered due to a temporary 3g restriction that had been imposed on the Javelin FAW.7 which virtually precluded quarter attacks. During September an interesting exercise took place in which the Hunter F.6s of 65 Squadron (also based at Duxford) acted as bombers with 64's Javelins as fighters. It was hoped that this would provide their day fighter compatriots with a better appreciation (and hopefully more respect) of the night-fighter in action. Battle of Britain day commitments saw 64 Squadron Javelins participate at the displays at Marham, Bassingbourn and Upwood and one aircraft acted as formation 'whip' for a flypast over Horse Guards Parade on the 20th. This month also saw the arrival of a Javelin T.3 dual-control trainer.

The rest of the year was taken up with a regular series of exercises (Ciano, Halyard, Kingpin) with aircraft being scrambled against bomber targets, mainly Valiants and Canberras. Good results were achieved as regards the number of 'splashes' and also when it came to scramble times, especially in view of the 'panel off' starting restrictions that were in place. This was due to the Javelin's reputation for start-up fires as the starter cartridges were mounted in a bay in the underside of the fuselage which was often contaminated by hydraulic and engine oils that had accumulated there from leaking pipes. For a time an inspection panel was left off on start-up so that the ground crew could check to see if anything was amiss. If all was well it was then replaced. Two Javelins took part in route training from 16–20 November when they flew to Nicosia via Orange, Luqa and El Adem, returning by the same route.

This trip was the precursor for another full squadron detachment to Nicosia when eleven Javelins took off from Duxford on 5 March 1960, arriving in Cyprus the following day having night stopped at Luqa. The route servicing party followed in a Beverley transport aircraft with the rest of the squadron having a 'very pleasant' direct trip in a Britannia. It had been decided beforehand that the planned system of flying would involve eight day and four night sorties and this was adhered to during the squadron's stay. Most of the flying was devoted to GCI and mutual PI's, many of the latter being flown at low level. A wide variety of aircraft were intercepted as they approached Cyprus and ranged from Valiants and Canberras to Hunters, Meteors, Skywarriors, Crusaders and Israeli Air Force Sud Vautours. The squadron also investigated a number of propeller-driven aircraft including DC-6Bs, Constellations, Britannias and Shackletons. Throughout the detachment a battle flight was maintained from dawn to dusk which led to nine scrambles and a number of aircraft were also diverted from normal PI sorties to investigate X-raids. Many personnel were also flown in helicopters during the squadron's stay in Cyprus and given experience of dry and wet, single and double lifts. The return to Duxford was made in one day on 6 April with the Javelins staging through El Adem, Luqa and Orange. A record was claimed for Javelin operations with the eleven aircraft achieving forty-four sorties in the day and the squadron Operations Record Book also took pride in the fact that eight of the Javelins were fully serviceable on their return to Duxford.

The rest of the year provided a regular schedule of exercises which included Fighter Command's annual summer exercise (Yeoman) from 19–23 May. This month also saw the departure of W/C J.A. Wright to a posting at the Air Ministry, his replacement as OC being W/C W.H. Mills DFC. No. 64 Squadron had a 'most memorable day' on 6 July when it was presented with a squadron standard in a joint ceremony at Duxford with 65 Squadron. The standards were presented by Marshal of the RAF Sir William Dickson GCB KBE DSO AFC and a combined squadron flypast was led by F/L P.G. Mallett, 64 Squadron's deputy 'A' Flight commander. Towards the end of the month the squadron began to re-equip with the Javelin FAW.9. Deliveries were rather slow and although seven FAW.7s were flown back to the Gloster airfield at Moreton Valence, only three FAW.9s made it the other way.

September was a busy month with a visit by crews of the All-Weather Fighter Combat School on the 13th. They flew on a number of training sorties and afterwards interception techniques and squadron procedures were thoroughly discussed. A further visitation was made towards the end of the month by Instrument Rating examiners who arrived in a Javelin T.3. Relations became rather strained when a

64 Squadron pilot 'blew up' their T.3 when he was starting it
and thoughts of a conspiracy became more deeply entrenched
when the squadron's own T.3 went unserviceable shortly
afterwards. They were placated to a certain extent by a few
trips in the squadron's FAW.9s on PI's and finally left looking 'somewhat appeased'. The month also saw the arrival of
S/L D.J Lomas to become 'B' Flight commander and a first
for F/L B. Smith who somehow managed to intercept a
Blackburn NA.39 (Buccaneer) during Exercise Flashback on
the 21st.

By the end of October the squadron had fourteen Javelin
FAW.9s but seven of these were unserviceable with faulty
booster pumps. These items supplied fuel to the engines and
this particularly snag was to be a recurring theme of Javelin
operations for some time. Aircraft availability was not helped
on the 25th when F/L E. Clarke was informed that there
were sparks coming from the starboard engine of the T.3
that he was flying. He landed without further incident but
on inspection it was found that the engine had stripped the
first-stage rotor blades and had damaged the second-stage
as well. An unusual trial took place on the 27th when F/L
K.A. Pye carried out radar controlled evasive action in
an air-miss with a de Havilland Comet. On the two sorties
that were flown rate ½ turns were initiated by the ground
control radar at 30 miles head-on separation, firstly for the
Comet, then for the Javelin and finally with both turning.
The fact that this took place the day after the Queen's Comet
had been 'intercepted' by *Luftwaffe* F-86 Sabres was felt to be
mere coincidence.

The rest of the year was dominated by preparations for the
introduction of in-flight refuelling and the use of Firestreak
air-to-air missiles, two developments that would transform
the Javelin. In-flight refuelling had been introduced on the
Javelin by 23 Squadron and W/C Chapman and S/L Jeffries
from that unit gave lectures on the new technique which
were extremely informative. Squadron members also paid
two visits to the de Havilland factory at Hatfield where they

learned all about Firestreak theory and practice. The first Firestreak acquisition sorties were flown in December by F/L B. Smith and F/L Don Headley and were completely successful. Also during December the first eight crews selected to train for in-flight refuelling were despatched to Marham to meet the Valiant crews with whom they would be operating and to attend lectures on techniques and rendezvous procedures. In the meantime ground crews were busy fitting refuelling probes and larger oxygen bottles to the aircraft at the rate of two aircraft per day.

The first dry 'hook-ups' were carried out on 15 December by W/C Mills (with S/L J.S. MacRostie) and S/L V.J. Morgan (with F/L Ovenell), the former achieving twelve at his first attempt. The squadron had been loaned the services of F/L Howard Fitzer of 23 Squadron and he flew a 'chase' aircraft to offer helpful advice from the sidelines. The next day, five more crews tried their hand at probing, again with F/L Fitzer flying 'chase'. It was not just the pilots who were working flat out however as the navigators were also extremely busy with logs and charts, carrying out limited aids cross-countries in preparation for future long-range trips to parts of the world lacking in modern navigation aids.

The in-flight refuelling training programme was continued in January 1961 and progressed to wet 'hook-ups' where fuel was taken on board. This allowed a series of medium-range cross country flights to be made with flight times of up to 3¼ hours. Such prolonged trips at altitude caused technical problems however as a lack of winterised seals for the undercarriage oleo legs led to a number of collapsed oleos, a situation that was not helped by poor spares backing. The unserviceability rate averaged 25 per cent for the month so that the flying task was not met. Exercise Razors Edge took place on the 3 January with one (of three) Javelins claiming two F-100 Super Sabres and the Firestreak programme also moved forward with exercises incorporated into normal PI sorties.

During February twenty-nine Firestreak acquisition exercises were flown and cross country flights with in-flight refuelling were extended up to 5 hours. Poor serviceability still caused problems for the ground crews during the month with several aircraft AOG (aircraft on ground) for booster pumps. To try to alleviate the situation two aircraft that had been damaged Cat.3 were turned into 'Christmas trees' and were robbed for spares as required to keep some of the other Javelins in the air. Serviceability was made even worse by a number of failures of the Mark 12 Horizon Gyro Unit (cockpit attitude indicator) and S/L V.J. Morgan who was one of those to experience a failure in flight was sent to the manufacturer (Sperry) to discuss the causes and possible solutions.

The month of March was largely given over to preparations for another squadron detachment to Cyprus and this included the fitting of radio compasses and Rebecca in each aircraft in place of the normal Gee equipment. On the 27th five Javelins arrived from Coltishall, each bringing four operational Firestreak missiles which were to be taken to Cyprus. The first to depart was Red section (including the T.3) which flew out in stages on 4–5 April to be followed by Blue section led by W/C Mills using in-flight refuelling on the 6th. A rendezvous was made with three Valiant tankers from Marham but only after a tailchase, and with a lower operating height than expected, plus the carriage of Firestreak missiles, each aircraft had to take on approximately 1,200 lbs more fuel than anticipated at the first refuelling. Because of this the second refuelling was brought forward by six minutes to allow the tanker adequate reserves of fuel. The second refuel was completed and then a rendezvous was made with tankers operating out of Idris on the north coast of Sicily. After successfully completing the third and fourth refuels, the last of which was just a top up, the section continued independently to Nicosia, landing after 5 hours 5 minutes in the air. Green section flew out on the 7th, also using in-flight refuelling (within two hours of arrival on the

5th, one aircraft of Red section had been placed on Battle Flight standby).

Air-to-air firing took place during May but on the 4th F/O R. Johns had his elevator jam as he applied roll in preparation for another attack. By using full forward stick and trimmer he managed to maintain level flight with an indicated airspeed of around 210 knots. He found that he had elevator control down to about 165 knots and carried out a successful landing at Akrotiri by flying a GCA with full forward trim and airbrakes to control the rate of descent. An inspection of the controls after landing found a screw jammed in the elevator linkage. During the month the flying target was not met mainly due to aircraft being AOG for booster pumps and starter units. This situation was not helped by a strike at the manufacturers of the booster pumps. Worse was to follow on the 27th when Javelin 'Lima' was started for a port engine run and part of the turbine disintegrated. The resulting explosion caused Cat.4 damage and bits of white hot turbine blade started a fire in the grass behind the ASP.

The main concern in June was to get all the aircraft serviceable as the squadron was scheduled to return to the UK on the 21st. Practice interceptions were continued until a week before departure and then the emphasis was on preparing the aircraft to leave. Such was the inadequate supply of booster pumps however that it was only by robbing aircraft left behind at Duxford that the squadron managed to complete its serviceability in Cyprus in time for the fly out. Even so three aircraft had to be left behind, the aforementioned 'L', plus 'P' which was Cat.3 with burnt electrical looms following a fire during engine initiation and 'C' which was awaiting an engine change. The return trip was not without incident as the Javelin of F/O R.J. Wark had a battery problem at Pisa and had to be left behind, while S/L D.J. Lomas had a double flame out at altitude. Both booster pumps indicated failure but the generators and hydraulic pumps continued to work. The aircraft descended rapidly but on re-selecting the booster pumps they indicated serviceable. At 25,000 ft the port engine was successfully relit but the starboard engine could not be re-started, despite numerous attempts. With 80 per cent set on the port engine the aircraft descended to 12,000 ft where it maintained height before an incident-free landing was made at Orange. The remaining aircraft landed at Wattisham to have their Firestreak missiles removed for storage and then all except one (which had starter trouble) flew back to Duxford.

Also in June three crews from 64 Squadron joined with five crews from 23 Squadron for Exercise Pounce which was a long-range deployment of Javelins to Karachi in Pakistan. The outbound flight was accomplished with in-flight refuelling and night stops at Akrotiri and Bahrain and the Javelins spent ten days in Karachi before returning via the same route. The development of rapid long-range deployments came just in time and the beginning of July saw 64 Squadron on readiness due to the situation in the Middle East where President Qasim of Iraq had claimed Kuwait. The Amir of Kuwait appealed for help from the UK and Saudi Arabia and the British response was to send a naval task force backed up by a squadron of Hunters. On 4 July five Javelins of 64 Squadron were requested to reinforce 23 Squadron and these were flown to Coltishall in the afternoon, leaving for Nicosia the following day. Whilst in Cyprus they flew normal PI sorties and shared the standby commitment with 23 Squadron. They returned to the UK on the 31st returning to Waterbeach where the squadron had moved to on the 13th. The move was popular, the squadron ORB stating that it felt 'like moving from a bed sitter to a mansion ... enough room for everyone's Empires to flourish ... even the technical chaps seem happy'.

The move to Waterbeach did however see many of 64's long-established crews depart to be replaced by crews fresh out of OCU. This meant that they had to be trained as regards flight refuelling techniques and this was accomplished over the next four months. The highlight of September was the

Battle of Britain display at Waterbeach on the 16th which was attended by a crowd estimated at 50,000. They were treated to a display by 92 Squadron's aerobatic team, the Blue Diamonds, and 64 Squadron was represented by F/L Don Headley who gave a handling demonstration of the Javelin. By now the training commitment had been extended to include a greater emphasis on interceptions of aircraft in conditions of ECM and also bomber affiliation. The latter involved an initial interception after which the main purpose was to give the Javelin pilot gunnery training in following an evading V-bomber target, an exercise that proved to be popular all round.

The last few crews completed their in-flight refuelling training in November during which W/C Mills showed how it should be done with a series of 'hook-ups', with the tanker winding in and retracting the hose after every eight contacts. A total of forty-three 'hook-ups' were completed in the hour which was claimed as a 'whip it in and whip it out' record. On 27 November another Javelin was lost when it was being flown by S/L Saunders-Davies who was attached to the squadron prior to taking command of 85 Squadron. While acting as a fighter on a night PI exercise the pilot noticed a lightening of the controls in the pitching plane during a turn to starboard and shortly afterwards the No. 3 hydraulic warning light flickered before going out. The exercise was quickly discontinued but when overhead Waterbeach and turning onto the outbound heading, the No. 2 hydraulic warning light came on and within a few seconds was joined by the No. 3. Realising that ejection was inevitable S/L Saunders-Davies warned his navigator (F/L P. Daughtry) and straightened the aircraft onto a southerly heading towards open country. Both crew ejected safely and the Javelin (XH878) crashed in a field at Fox Farm, Balsham, ten miles south-east of Cambridge.

There was another emergency in December when F/L A.L. Button in XH871, after refuelling from a tanker during a cross-country exercise, found that aileron control was limited to 30 degrees of bank. He diverted to Leuchars during which the aileron control gradually deteriorated and when still twelve miles from touchdown on a GCA approach it became impossible to bank the aircraft by aileron. Rudder was used to control the aircraft during the rest of the approach and a safe landing was made. The post-flight examination showed that the aileron lines were fouled against the hot air ducts and pieces of foreign matter were also discovered in the proximity of the runners.

Much of December and the beginning of the New Year was beset by bad weather, initially fog, but eventually snow arrived and Waterbeach received a 10 inch covering. The first Exercise of 1962 was Flodden from 5–11 January and two Javelins flight refuelled to Cyprus on the 22nd before staging back through El Adem, Luqa, Pisa and Orange by the end of the month. In February W/C Mills and two other Javelins flew to Decimomannu in Sardinia to inspect the facilities there with a view to using the airfield as a staging post for future operations. Back at Waterbeach a Javelin sustained Cat.3 damage in the air when the starboard rail of the rear hood became detached, breaking the Perspex and damaging the airframe as it departed. It was later found in a field six miles south-east of the airfield.

By now serviceability was much improved as the main snags likely to affect the Javelin FAW.9 were well known and ground crews had little difficulty in keeping enough aircraft on line to complete the flying task. Indeed the main problem appeared to be maintaining enough crews as, out of an establishment of eighteen, the effective strength in March was only sixteen with little hope of any replacements. As the squadron was now getting a reasonable ration of ECM targets during PI's the navigators were now much more confident with this type of interception. There was also further experimentation with flight refuelling techniques when S/L V.J. Morgan performed dry 'hook-ups' with a Sea Vixen of the Fleet Air Arm. During the month the squadron said goodbye to another commanding officer when W/C Mills was replaced by W/C J.S.W. Bell.

Anyone who thought that all Javelins on 64 Squadron coded 'L' were jinxed had more reason to think this way after 13 April, which also happened to be a Friday. FAW.9 XH844, which was the replacement for the 'Lima' that had exploded on start up in Cyprus the previous year, suffered a similar fate when there was an explosion as the port engine was being started. The starboard engine had been started successfully and was running normally but the drama began as soon as the port engine starter button was pressed. Neither of the crew was seriously injured, though the rear cockpit was blown in from both sides causing minor injuries to the navigator. The damage was sufficient for the Javelin to be declared Cat.5. The subsequent investigation showed that a clamp on the exhaust pipe had become loose and this had allowed hot gases to escape into the servicing bay. The gases had then mixed with hydraulic fluid mist to produce a mixture that was very close to its flash point. This was reached with further rise in temperature as the port engine started up causing the mixture to detonate. There was another start-up fire the following month involving XH765 but in this case damage was not as severe and was assessed as Cat.3.

There were further problems when XH872 caught fire on start-up on 14 June. When attempting to start the port engine the rpm rose to only 10 per cent before decreasing again. The starboard engine was then started normally and after about three minutes a second attempt was made to start the port engine. Frantic waves from the ground crew alerted the pilot to the fact that something was amiss and he quickly shut down all the aircraft's systems. Yet another fire had taken hold in the servicing bay and electrical cables were badly damaged before it could be put out. Also in June, XH764 had to be declared Cat.3 with a cracked leading edge. With a shortage of aircraft and an overall deterioration in the serviceability rate, the flying task was not achieved for the month. There were enough aircraft however to take part in various exercises in June including a Razors Edge on the 13th during which F/L K.A. Pye and F/L Don Headley had considerable success, claiming a total of ten F-100 Super Sabres between them.

Most of the excitement for July was taken up by the squadron's first missile practice camp (MPC) at Valley. Before firing, each of the five crews that had been selected had to complete a sector recce and two practice range firing sorties. The aim was for each crew to fire one live missile and the results were rather mixed. The only direct hit was the missile fired by F/L Don Headley and Sgt Eddie Buckingham, although there were near misses for S/L Lomas and W/C Bell, and F/L Scott and F/L Scotton. The missile fired by F/L R.L. Holmes and S/L P. Gilliatt broke lock soon after firing and the Firestreak fired by F/L R.L. Lockhart and F/O Deane appeared to have a mind of its own as it made no attempt to follow the target and disappeared into the blue. There was yet another start-up accident in July when the port starter injector disintegrated on XH871 and caused extensive damage.

Having become used to the relative luxury of Waterbeach it was all change again in August when the squadron moved to Binbrook in Lincolnshire on the 22nd. The last three weeks at Waterbeach were relatively quiet with an Exercise Flinders from 7–14th and a Kingpin Adex on the 16th after which flying was suspended until the move to Binbrook. Also during the month three of the squadron's FAW.9s departed for modifications to allow the fitting of long-range fuel tanks and the rest were to follow in batches over the next few weeks. To make up for the loss of these aircraft six Javelin FAW.7s were delivered as temporary replacements. During the month of September the squadron took part in Exercise Matador II during which it operated a three-flight system which gave crews twelve hours on and six hours off. Bad weather affected some scrambles and two aircraft had to divert to Leconfield and one to Middleton St George.

In October the last batch of FAW.9s left for conversion so that squadron strength comprised the Javelin FAW.7s and a single T.3. There was a gradual realisation that the weather factor at Binbrook was not what the squadron had been used to and day and night flying was curtailed on a number of occasions owing to poor visibility and the airfield's lack of radar approach aids. This, together with QRA duties which were taken over on the 21st, severely limited the squadron's night flying potential. November was no better and 50 per cent of the weather observations were of a visibility of 3,000 metres or less, however it was noted that this would have been enough if PAR had been available [calibration of the radar was expected to be finished before the end of the year]. QRA commitments were maintained throughout the month and included one scramble which intercepted a Varsity flying at 1,500 ft and 150 knots. Such slow speed caused the crew (F/L I. Rixon and F/L G. Pearce) some difficulty but a successful identification was made.

The weather deteriorated significantly in December and was noted as being the worst flying month for two years. A combination of fog, low cloud and rain, strong crosswinds, then severe frost and ice followed by snow decimated the flying programme so that it could only run on ten days and five nights. By the end of the month all but five of the FAW.9s had returned after modification and the FAW.7s had all been flown out [No. 64 Squadron continued to operate from Binbrook until September 1964 when one Flight of Javelins flew to Tengah in Singapore. Over the next six months this section was gradually built up to full squadron strength and the unit remained in the Far East until it was disbanded on 16 June 1967].

CHAPTER 15

Crew Debrief – 1

In early 1956 F/Sgt Bob Hillard was a member of 187 Squadron which was a ferry unit that was tasked with delivering aircraft from Maintenance Units to the squadrons and vice versa. Having flown Meteors and Vampires in the Middle East with 249 Squadron, he already had ten years flying experience by the time that he was introduced to the Javelin

I do feel that a fighter that is not to be looped is missing something. Even the Meteor NF could and they didn't look as though they should. Looking through Pilots' Notes again I am not sure that anyone should have been sent aloft in the beast. Apart from the stalling/spinning section, the page on warning lights and indicators is quite a handful of information. The same applies to the page on hydraulic failure and its effect on controls. Trying to work out which of the hydraulic pumps may have failed reminds me of a pools perm. Nothing was simple in the Javelin – even starting took up a lot of space in Pilot's Notes. I am sure that I never went through all of the checks shown and I bet the squadron chaps probably cut a few corners too. Now as far as we ferry pilots were concerned (who were not likely to be in a position of being near the stall, apart from the approach) you couldn't have had an easier aircraft to fly. On landing, the huge wing provided sufficient hold-off and often it had to be put on the runway. But there is always a catch. Nice as it was in the air, it was on the ground that I found problems. As an ex-Harvard pilot, the toe brakes should have been straightforward, but I don't think I ever really mastered keeping the thing in a straight line. I thought it was just me, but I noticed later when delivering Javelins to squadrons that already had them that there were a number of swerving skid marks on the runway, so it seems that even those who flew them regularly may have had similar problems.

My Javelin course was at Benson in January 1956 when I flew XA622 for six sorties totalling five and a half hours, my navigator, F/Sgt Gray Murrin, joining me for the last four sorties. Rank, it is said, has its privileges. We were only allowed to fly one 'modern' aircraft and with increased complexity, this made good sense. The Hunter, being a popular aircraft, did not get much of a look-in as far as the NCO pilots were concerned. As I couldn't see a way in to Hunter ferrying, as soon as there was a hint of 187 Squadron ferrying the Javelin I was quickly in with my application, which was accepted. At this stage there were all sorts of rumours about the Javelin but little of it was complimentary. Gray Murrin usually flew with me and we delivered the first Javelin (XA570) to 46 Squadron at Odiham on 24 February 1956.

It was deemed necessary to send me back to Benson for a short refresher course before ferrying the Javelin FAW.4. The boss of the FTU (S/L Hunter) first of all checked me out on a Meteor T.7 before deciding to fly with me in the FAW.4. I hadn't expected a dual ride, there being no controls, of course, in the navigator's position. The boss had his own ideas however and brave man that he was told me to go up to 40,000 ft and pull into a steep turn up to the judder. I performed for the brave gentlemen in the rear seat and relaxed the back pressure on the stick as soon as there was the first whiff of judder which seemed to satisfy him. I then flew back to Aston Down where I decided to do some steep turns over the accommodation we were in at the time in the hope that my wife, who was heavily pregnant, would come out and wave. I was rather disappointed when she didn't. At the time I was doing my bit it seems my wife was doing hers and was in the process of producing our second daughter who was born, as seemed appropriate, to the sound of two Sapphire engines.

On 3 May 1957 two Javelin FAW.5s were to be flown to Leuchars and F/Sgt Doc Watson (navigator F/Sgt Tiger Anslow) and I were to go in loose formation. We set off from St Athan with Doc leading and eventually arrived overhead at Leuchars. This was another first Javelin delivery so there was the usual crowd to see us in. For some reason we did a proper fighter break with me leading and I taxied in and climbed down, followed by Doc. It seems the drinks had been lined up in the Officers' Mess but the sight of four Flight Sergeants put an end to that. One young navigator approached Tiger Anslow and asked him what he thought of the Javelin. Now Tiger (so called because of his liking for Tiger Balm) was ex-Transport Command and believed that there should always be more than one engine, four if possible, and that these must always be props. Jet aircraft, or so he thought, were the work of Satan and just because he had been told to navigate one, didn't mean that he actually liked it. His reply was predictable and simple 'Bloody awful!' We made our way to the Sergeants Mess (no drinks lined up) where we were told that we were too late for lunch but there was some cold soup if we would like some.

F/O Tony Warner was one of the first squadron pilots to fly the Javelin during his tour with 46 Squadron at Odiham

The test flying programme was never completed regarding the stall and spin and, after two crashes on test, the Javelin was given to 46 Squadron. Since there were no dual seat versions available at the time, our conversion consisted of a flight in the navigator's seat and being shown how far we could go into the judder as the FAW.1 had no stall warner. I don't think this inhibited or restricted our use of the flight envelope in its role as an all-weather fighter. After all very few Meteor pilots had ever spun a Meteor. The Javelin was very easy to fly and was a vast improvement on the Meteor NF.11 with better speed, a better rate of climb and far greater radar range and better reliability. It could be difficult to land if speed was incorrect (usually in bad weather) and if airspeed was high the aircraft would 'bounce'. This could be difficult to correct due to slow engine response to full power. Engine start-up could be interesting. Generally if the cartridge fired normally the engine would start. Difficulties arose when the first cartridge didn't fire and the second cartridge was used. On returning the ground crew would replace the one that had fired, leaving the other in situ. The next crew would do exactly the same so eventually the first cartridge had been 'cooked' many times until firing of the second cartridge ignited both together, causing the starter to disintegrate. As a result of this only one cartridge was kept live, leaving an already fired cartridge in the other holder.

Having flown Meteor NF.11s with 68 Squadron in Germany, F/L Howard Fitzer also flew the Javelin FAW.1 with 46

Squadron at Odiham. He later flew with 23 Squadron when this unit introduced in-flight refuelling to the Javelin force and was involved with various trials to perfect the technique as he explains

As the first RAF fighter to be designed from the outset for all-weather operations (as opposed to modified day fighters), the prospect of a greatly improved perform-ance over the Meteor and the radical delta-wing design inherent in the Javelin presented an eagerly-awaited prospect to 46 Squadron. I recall an unconfirmed report that 46 Squadron was chosen to be the first squadron to receive the Javelin on its introduction into service on no more substantial a qualification than the design of its emblem (three linked javelins). Bearing in mind the propensities of senior management (in any sphere) I would not be the least bit surprised if the story was based on fact! Despite the lack of a two-seat trainer, transition to the aircraft provided no major problems. It possessed a roomy, comfortable cockpit and was easy to fly.

I did, however, have some reservations about the potential effectiveness of the aircraft under combat con-ditions, particularly its observed behaviour at incipient or full stall. I had experience of stalls and spins in the Meteor and had complete confidence in the aircraft's ability to recover. The prospect that recovery from these conditions (particularly in cloud or at night) may not be possible in the Javelin did not inspire confidence. Bearing in mind its 'all-weather' role, I was also con-cerned that poor collimator design robbed the aircraft of a capacity for blind gun firing using radar-derived target positional information in conjunction with the conventional gunsight. Perhaps the capacity to carry missiles would have made all the difference, particularly if collimator accuracy and reliability proved sufficient for missile target acquisition. In common with all wing-mounted guns, projectile 'pattern spread tended towards poor results, particularly when wing flexing under high 'g' loads led to inaccuracies in sighting (this was also a particular problem with NF Meteors). In my opinion the Javelin FAW.1 inspired affection rather than confidence. It was underpowered, built like a tank and flew like one!

On 23 Squadron I was heavily involved with the in-flight refuelling programme. After trailing of the drogue, a cautious final approach was required (at approximately slow walking pace) to allow for late corrections, especially in turbulence. Once the probe was located within the outer confines of the drogue, the latter's flexibility and conical shape allowed it to accommodate minor positional errors prior to locking. Within trail limits, it was an advantage to have as short a hose trail as possible, the length of which appeared to some extent to be responsive to the fighter's throttle setting. In calm conditions the drogue trailed in a reasonably stable manner, and only a little concentration on smooth, accurate flying was required to effect a successful hook-up. In turbulent conditions, however, it was a very different story. Apart from the difficulty of achieving initial hook-up, any subsequent undue movement of the fighter in any plane relative to the drogue, particularly if the drogue itself was unstable, tended to induce a 'snake-like' movement in the hose. What initially appeared to be a tenable oscillation could, by a process of compounded fighter pilot corrections and drogue instability, become a violent 'nozzle-snapping waggle'. If the side loading thus induced exceeded 1,000 lbs then the probe nozzle was designed to fail (i.e. snap off) in order to avoid structural damage to the aircraft.

Unfortunately a 'snapped-off' nozzle would remain firmly locked in the drogue, thus preventing further use of that drogue. Clearly a problem could arise if during a particular operation several fighters required virtually

152

elin T.3 XH391 flew with the RAF's Handling Squadron and later served with 228 OCU before
ng scrapped in 1964. (*Philip Jarrett*)

ormation of missile-armed Javelins led by FAW.7 XH908. The other aircraft comprise FAW.9s
I768, XH881 and XH882. (*Philip Jarrett*)

Javelin FAW.7 XH778 of 23 Squadron was later converted to FAW.9 standard and flew with 29 Squadron. It was struck off charge in August 1964. (*Howard Fitzer*)

Javelin FAW.9 XH881 'M' of 25 Squadron. (*Philip Jarrett*)

A number of Javelin FAW.9s were converted to '9R standard which allowed four 230-gallon fuel tanks to be carried. XH887 of 64 Squadron carries two such tanks on the inboard pylons with Firestreak missiles outboard. (*Philip Jarrett*)

A Firestreak missile on its trolley is prepared for fitment to Javelin FAW.9 XH880 'J' of 25 Squadro This aircraft carries the initials of C.O. W/C J.H. Walton on the fin. (*Philip Jarrett*)

Another view of Javelin FAW.9 XH965. This aircraft did not serve with the RAF and was used exclusively on trials work. It is seen here at RAE Bedford. (*Philip Jarrett*)

...elin FAW.7 XH778 of 23
...uadron in the high-drag
...nfiguration with
...dercarriage down and
... brakes deployed. (*Howard Fitzer*)

...view from the cockpit of a
...Squadron Javelin FAW.9 as it
...ses on the drogue trailed by a
...liant B(K).1 of 214 Squadron
...ring the first in-flight
...uelling trials in 1960.
...*ke Miller*)

A Javelin FAW.9 of 23 Squadron takes on fuel from Valiant B(K).1 XD812 of 214 Squadron. (*Mike Miller*)

Javelin FAW.9R XH891 of 29 Squadron fitted with four 230-gallon overload fuel tanks. (*Philip Jarrett*)

...is view of a Javelin FAW.8 taking off shows to advantage the reheat nozzles for the Sapphire ...7R engines and the vortex generators ahead of the ailerons. (*Philip Jarrett*)

...velin FAW.7 XH758 of 23 Squadron. It was later converted to an FAW.9 but was lost on ... October 1963 when the starboard engine blew up as a result of centre-line closure whilst serving ...th 5 Squadron. (*Philip Jarrett*)

Ground servicing taking place at RAF Luqa in Malta in October 1966. Prominent in this view is the rather cumbersome cockpit access ladder. *(Philip Jarrett)*

Javelin FAW.9 XH898 of 228 OCU at Leuchars in a natural metal finish. It carries the initials of S/L G.H. Beaton on the fin. *(Philip Jarrett)*

Ground crews toil to change the starboard main wheel of a Javelin FAW.9. (*Philip Jarrett*)

Javelin FAW.9R XH876 survived until 25 August 1966 when it lost power and crashed on approach to Tengah whilst serving with 64 Squadron. (*Philip Jarrett*)

Like a number of other aircraft of the period, Javelin FAW.9R XH834 of 64 Squadron has the initial of its C.O., W/C P.D. Wright, on the fin. *(Philip Jarrett)*

A general view of the apron at Ndola Airport in Zambia shortly after the arrival of 29 Squadron's Javelin FAW.9s on 3 December 1965. *(Philip Jarrett)*

velin FAW.9 XH725 'P' of 60 Squadron. This aircraft had previously flown with 29 and
Squadrons and was eventually struck off charge on 1 May 1968. (*Mike Miller*)

aving initially flown with 23 Squadron as an FAW.7, XH777 was converted to an FAW.9 and was
.en used by 29 Squadron. It then served with 60 Squadron (seen here) until a belly landing at
engah on 5 February 1968. (*Mike Miller*)

Two Javelin FAW.9s of 60 Squadron. The tail markings comprise black and white bands and the head of a markhor, a wild goat found in the Khyber Pass in India, a region the squadron operated over for a number of years. (Colin Edwards)

A view of Javelin FAW.9 XH719 as it flies in close formation with another 60 Squadron Javelin. *(Philip Jarrett)*

In addition to the normal tail markings, Javelin FAW.9 XH960 of 60 Squadron displays a lightning flash on the outer wing pylon, a marking previously applied to 60 Squadron's Vampires and Venoms. *(Philip Jarrett)*

e disbandment of 64 Squadron at Tengah on 10 June 1967 was marked by a flypast of nineteen
velins, comprising an arrowhead of ten aircraft from 64 Squadron with a diamond nine from
Squadron. (*Philip Jarrett*)

The last airworthy Javelin was FAW.9 XH897 which flew with A&AEE at Boscombe Down in a red and white colour scheme. It was retired on 24 January 1975 and flown to Duxford to join the Imperial War Museum collection. (T. Malcolm)

simultaneous refuelling in the presence of disabled nozzles and/or drogues. The problem of turbulence was highlighted during a trial in which I was required to attempt in-flight refuelling at low level (about 2,000 ft) in the Welsh hills in mid-Summer. In the presence of strong turbulence I finally succeeded (after eleven attempts) in hooking up to the tanker, only for a 'snake' to develop in the drogue, sufficiently severe as to induce nozzle failure and a lateral collision between the drogue and the nose of the aircraft, the latter being slightly damaged. The trial was abandoned and as far as I know was never attempted again. This general problem in the presence of turbulence was also aggravated by the extreme length of the Javelin probe (20 ft) which, of course, in itself amplified (at the tip) any angular displacement of the aircraft.

When I came to fly the FAW.9 the sole advantage of reheat appeared to be improved performance at altitude, especially maintenance of speed in the turn with approximately 12 per cent extra power. The disadvantages were a loss of performance below 20,000 ft when reheat was engaged. As I understood it this occurred because combined fuel demand of main engines and reheat exceeded available fuel-pump capacity. Since the reheat was less efficient than main engines for a given fuel flow, below 20,000 ft power was lost because the reheat system was allocated priority fuel supply. The practical indication of power loss below 20,000 ft was a 5–10 per cent drop in engine rpm when reheat was engaged. The reheat exhaust gases were highly visible at night (even from ahead), posing a problem of being spotted by a potential target. There was also a high level of unserviceability as one or both engine reheat systems would frequently fail to function.

By the late 1950s F/L Noel Davies was an extremely experienced night-fighter pilot having flown the Mosquito NF.36

with 39 Squadron at Kabrit in Egypt, the Vampire NF.10 and Venom NF.2 with 23 Squadron at Horsham St Faith/ Coltishall and the Meteor NF.11 with 68 Squadron at Wahn as part of 2 ATAF in Germany. In June 1957 he was posted to the All-Weather Development Squadron at West Raynham where he flew all marks of Javelin up to and including the FAW.7

Javelin flying was something quite different from what I had been used to. The hydraulic wheel brakes, excellent power, superb hydraulic controls, good range and a ceiling of around 50,000 ft (gun-armed FAW.7). It was also supersonic if you were not careful and it had the most dramatic airbrakes. I gave a demonstration flight to a civilian (probably a Ministry of Defence man) and coming back into the circuit under full power at about 400 knots, which was the maximum speed at which the airbrakes were allowed to open fully, I opened the air-brakes. It was like hitting a brick wall. The airbrakes were about 2 ft by 8 ft on top and below each wing, and came out almost at right angles. Needless to say there was a grunt in the back as my guest lurched forward and hit his (helmeted) head on the coaming, so I apologised and put them in again, at which point the full power took hold and he banged his head on the headrest – hard! It was a most graphic demonstration however uncomfortable for the uninitiated.

When I arrived at AWDS there was no Javelin trainer at that time and so we had a familiarisation trip in the back seat, read Pilot's Notes and off we went. Only eleven days after my first Javelin trip, on my tenth sortie, one of the engines 'threw a blade' which fortunately for us went outwards and not inwards. The effect was small and we thought that we had some minor hydraulic trouble (judging by the rumble) but we decided to return to base. Because of the wind we had to land on a short runway and fearing possible (hydraulic) brake trouble we went round twice to get everything exactly right for landing. On approaching dispersal we were given signals to cut the engine quickly which we did as smoke was coming out of the rear of the engine compartment. I think twenty-two blades had been damaged or were shed. We were very lucky and were amazed at the superb engine performance under those circumstances. Not long after F/L Arthur Wright and F/L Reg Ashworth (also from AWDS) had a similar fault over the Wash but the shed blades travelled inwards through the fuel gathering portion of the engine causing a fire from which the aircraft (and they) did not survive [they were flying FAW.5 XA642 on 6 December 1957]. Other similar incidents occurred including one that was detected at night because the flames were reflected from cloud on the climb out after take off. On this occasion the crew baled out successfully. The interim cure was to install a wire-mesh 'basket' arrangement round the compressor disc to prevent shed blades escaping to damage more vital parts. Proper corrective action was designed later.

Our job at AWDS was to investigate the performance envelope for squadron operations to make the best use of the aircraft in normal service terms. One particular aspect I remember is the assessment of the best way to return to base on minimum fuel. We investigated various profiles, including cutting one engine at height (48,000 ft), throttling the other back as far as possible, extending airbrakes and descending at maximum practicable speed. The effect was a very shallow initial descent, gradually increasing at a steady Mach number and then slowly becoming shallower again as we converted to IAS. This brought us down to circuit height in about 3½ minutes. The drawback was that the aircraft remained very cold and iced up badly, but the Javelin had such a superb windscreen defroster (hot air from the compressor which was blown on to the outside of the screen) that this was almost a non-problem in the event.

F/O Guy Woods had a long RAF career as a navigator on all-weather fighters. Having completed his radar training on the Bristol Brigand, he converted to the Meteor NF.12/14 at North Luffenham before joining 151 Squadron at Leuchars in 1956, just as the unit was in the process of swapping its Meteor NF.11s for Venom NF.3s. However it was not long before the squadron began to fly the Javelin and the new type was to make a favourable impression as he explains

I did some eighteen months on the Venom before 151 converted to Javelin FAW.5s. The squadron was on a long detachment to Turnhouse while they extended the runway at Leuchars so we converted to the new type there. This was done by the Javelin Mobile Training Unit, an outfit which went round the squadrons with mobile classrooms and a couple of Valetta aircraft fitted out with the new AI.17 for the navigators to get to grips with. As I recall the conversion took several weeks and for us the Javelin was a quantum leap. It was big, powerful (and comfortable) with a good radar, four 30 mm cannons and the ability to operate at high subsonic speeds well above 40,000 ft. Most importantly there were ejection seats for both crew members. It also had a good rate of climb taking six minutes to reach 40,000 ft and twelve minutes to reach 50,000 ft which was outstanding and compared favourably with the Hunter. The Javelin could turn quite well at these higher levels and you could complete an interception at 50,000 ft with a little care. It was marginally supersonic and was capable of about Mach 1.04 in a steep dive. I remember that questions were asked in the House of Commons during the 1950s regarding a supersonic boom being heard over London during the time when the politicians were having doubts about the Javelin's capabilities. The test pilot concerned was quoted as saying that he had accidentally slipped through the sound barrier while on a routine test flight. You could not accidentally slip through in a Javelin; you had to work very hard to do it!

The radar was quite nice to operate. It still had a two-scope display but the Range/Azimuth picture was in a B Scope format which was different from the PPI of AI.21 on the Venom. This took a bit of getting used to, but of the two formats I preferred the B Scope. It gives a very distorted view of the outside world but overall it is a much better display for interception work. A rough and ready azimuth/elevation display was piped through to the pilot's gunsight so he could follow evading targets. This took a load off the navigator but we still practiced the art of following evading targets from the back seat. As in AI.21 the radar was not roll stabilised so we were still left with the problem of scanner control during banking manoeuvres. In actual fact the radar was roll stabilised but the system was totally useless and was never used. However, we could lock the radar on to the target which took most of the heat out of Visual Ident runs. The designers of AI.17 miniaturised as much as they could, including the B Scope CRT. To make up for the deficiencies of the little tube they had to fit a high power magnifying glass over it so that you could see the display! Although we now had a match for most of the targets we might expect to come up against (ours as well as theirs) life and tactics went on much as before. Whereas with the Venom we had our work cut out to get up to something like the Canberra's operating height and had only a marginal overtake capability, we could now catch these quite easily. All the other squadron activities continued as before.

Towards the end of 1958 my pilot (F/L Ken Bassett) and I were posted as a crew to 25 Squadron at Waterbeach. This outfit was still equipped with Meteor NF.12/14s so we had two or three months back on Meteors before we started converting to the Javelin FAW.7. The mobile unit once again did the conversion but about a third of the squadron were newcomers from other Javelin squadrons, thus providing a nucleus of experienced crews. The Javelin FAW.7 was much the same as the FAW.5 but had the more powerful Sapphire engines and could carry four Firestreak infra-red guided missiles in addition to two 30 mm cannons. There were other internal differences, but operationally it was much the same, although the extra armament gave us a much increased ability to take out higher performance targets. The FAW.7 was thought by many to be the best Javelin of all in terms of overall performance and combat capabilities. The radar was much the same as AI.21 in terms of pick-up range i.e. 15–25 miles or so.

Our tactics were changed somewhat to meet the expected threat. We concentrated on interceptions of air-craft which were always assumed to be on a reciprocal heading to ours, simulating the Javelin force going out to meet a mass raid. The radar performance left us with little chance of adjusting the interception geometry at long range as we always did a 210 degree turn when the target reached a range of 12–13 miles. This catered for the target which was dead head-on and would put us behind the target at a good range with 30 degrees of convergence on it to complete the interception. Any displacement from targets track was a bonus allowing us to maintain radar contact with the target all the way round the final turn, whereas with little or no displace-ment, contact was lost for some of the time. Similarly, if the track-crossing angle was less than 180 degrees this proved to be a bonus in our favour.

All three V-Bombers were on line by now and they could cruise at a much higher level and faster than the Canberras, putting us back to square one again. How-ever Firestreak gave us a significant advantage over these targets. Air-to-air gunnery practice continued and we also now had the chance to fire the missiles at Jindivik drones. I did not get the chance to fire at one of these in my remaining time on the squadron and had to wait for many years before I eventually had a crack at one with a Sidewinder from an F-4 Phantom! Squadron life and operations continued as before with the threat still deemed to be at high level. After some months with the FAW.7s they were taken away from us one by one to be converted to FAW.9s, the main difference being the installation of an afterburning system for the Sapphire engines. This turned out to be useless below 20,000 ft, in fact you got a loss of thrust at sea level, but above 20,000 ft they certainly enhanced climb rate and improved your turning performance at high levels. However the squadron bosses didn't like us using reheat because it substantially reduced flying hours on an aircraft which didn't have much endurance in the first place. With the new mark we were able to complete interceptions at around 50,000 ft and indeed a little above, although the aircraft oxygen system supposedly limited us to 50,000 ft.

A year before I left the squadron I completed the All-Weather Fighter Combat Leaders' course at West Raynham with my flight commander and we explored a whole new area of radar fighter tactics. There was a lot of emphasis on air-to-air gunnery and all aspects of the Javelin's weapons system were studied. Interceptions at various levels were practiced including those at very low level over the sea against targets flying at a variety of speeds, something we did not normally do on a squadron. Of interest, we carried out these interceptions against low level targets flying at moderate speeds on one engine to conserve fuel. Against higher speed targets we used both engines but were limited to a maximum speed at low level of 535 knots indicated because of tailplane flutter problems, although the Javelin could go a lot faster. We were also taught and practised interceptions against jamming targets with both noise ECM and window/chaff. Specialist jamming Valiant bombers provided this service and we didn't normally get the chance to experience this during routine squadron training operations. The only time we would get to see both forms of jamming would be during major exercises and, of course, in a war situation. Interceptions could be successfully completed against noise jammers without too much trouble and you were only fooled by window the first time that you saw it! Life was a bit more varied on my second tour and we actually managed to deploy overseas, usually to Cyprus (both Nicosia and Akrotiri) and occasionally to Germany.

Overall my time on 25 Squadron was pretty quiet although there were a few incidents. On one occasion the starter motor fuel (AVPIN) had collected in the large equipment bay and caused a big explosion under the aircraft on start-up. This ripped the aircraft open from nose to tail and caused the hurried evacuation of the crew. The ladder was still in place so at least they didn't have a ten-foot drop to the ground to contend with! In those days we had very few 'career' officers on a squadron and those that we had were usually ex Cranwell cadets. One of these had an unfortunate experience while starting up a Javelin for flight. He hadn't noticed that the undercarriage 'Up' button had been selected and when the hydraulics came on line the gear started to retract. He managed to shut down before the main wheels started to move but this left an amusing sight of one Javelin in a line of about ten with its tail sitting high in the air. Needless to say the pilot's career suffered a set back.

The Javelin had some undesirable aerodynamics in that you were not allowed to stall or spin it or do any manoeuvres which could get you into this area of flight. Looping manoeuvres were specifically prohibited. Pilots did do them occasionally out of devilment, but not mine

I'm glad to say. I recall that an American exchange officer obtained special dispensation to do a loop prior to returning to the States. As luck would have it he stalled at the top of his loop, couldn't recover and crashed, killing himself and his navigator. My worst experience in a Javelin took place one dark night. We were flying along at about 45,000 ft when the Javelin started pitching and yawing quite violently – all symptoms of a stalled situation. We started dropping like a stone and I kept a beady eye on the altimeter and airspeed indicator reassuring my pilot that we were actually still flying. We lost about 10,000 ft before he realised that it was the yaw damper which was causing the rudder to travel violently from one end stop to the other! Other than this I experienced the occasional engine and hydraulic failures, together with other sundry malfunctions, but nothing particularly life threatening during my time with 25 Squadron. In general the Javelin was well liked in service; it performed well, it had a good radar and was very comfortable for both crew members. I certainly enjoyed my time on them.

Guy Woods left 25 Squadron in May 1961 to become a Launch Control Officer on Thor ballistic missiles but returned to flying duties after two years when he joined 31 Squadron on Canberras. He later spent time at A&AEE where he participated in flight trials and here he met up again with the well-known red and white Javelin FAW.9 XH897 which was part of the fleet at Boscombe Down, an aircraft he had flown on a number of occasions during his time with 25 Squadron. He later served with 29 Squadron on F-4 Phantoms at Coningsby and ended his career as a ground instructor on the Tornado F.3 simulator with 229 OCU.

F/O Ivan Logan was another Javelin navigator who had previously flown de Havilland Venom night-fighters. This aircraft was equipped with the American-designed

Westinghouse APS-57 radar (British designation AI.21) and was similar in its presentation of the radar picture to the AI.22 of the Javelin FAW.2, 6 and 8

> With the AI.21 on the Venom I remember picking up a USAF B-36 at a range of twenty-eight miles and usually on something like a Valiant we would get contact at about eighteen miles. Following evading targets at close range on AI.21 was not easy as our azimuth scope was a PPI. On AI.17 the time base was expanded right across the tube so that an aircraft on a collision course came straight down the screen rather than towards the centre of the time base [as on AI.21 and 22]. In effect with AI.17 the movement of a target at close range was exaggerated which therefore made it easier to follow. AI.17 also had a lock on facility. We usually locked on at about two miles and from there we only had to give the pilot the interception commentary without having to work all the radar controls. We also had a collimator which gave the pilot a presentation on his gunsight which he simply had to follow; all we did was to call out the ranges. The Javelin was ideal from the navigator's point of view being built as a night-fighter from the outset. It had plenty of room, had decent lights and most importantly of all as far as the crew was concerned 'bang seats'. Most of our Venom accidents were fatal [the Venom night-fighter was not fitted with ejection seats] and although quite a few Javelins were lost, there were a number of successful ejections.

Having flown the Javelin FAW.4 and FAW.7 with 23 Squadron at Horsham St Faith and Coltishall, Ivan Logan went on to become an instructor on FAW.5s with 228 OCU at Leeming.

Throughout this book there are a number of references to the American-designed AI.22 radar suffering from serviceability issues in RAF service. Although it was an extremely capable radar, this was also the experience of F/O Paul Hodgson who was a pilot with 89 Squadron

The Venom NF.3 was not my favourite aeroplane unlike the Javelin which followed it. No. 89 Squadron converted to the Javelin FAW.2 and FAW.6 at the end of 1957 and was subsequently re-numbered as 85 Squadron at Stradishall. In 1959 we moved to West Malling where we converted to the FAW.8. Over subsequent years I flew all marks of Javelin except the Mark 1 and 9R in the UK and Germany. I enjoyed my years on the Javelin and found it a very good aircraft to operate in the all-weather role and occasionally as a day fighter where it could surprise a few people. It was reliable, comfortable, an excellent instrument platform and carried quite a formidable armament.

The radars fitted (AI.17 and AI.22, depending on marks) differed considerably in serviceability. The AI.17 was more reliable for day-to-day operation but could not approach the performance of a well set up AI.22. Unfortunately the RAF never seemed to have the facilities to get the best from AI.22. While I was at West Malling an American team came for a week to instruct on the maintenance and setting up of the AI.22 radar and the results were impressive. We saw a level of performance that no one expected but, sadly, this improvement could not be maintained once they left.

Having started his operational career on de Havilland Mosquitos, F/L Bruce Spurr later became a QFI on Meteors and Vampires before flying the Javelin with 228 OCU at Leeming

The Javelin was an easy aircraft to fly within the normal flight envelope. Any vices could be associated with the stall, the deliberate practice of which was prohibited for squadron pilots. If inadvertently stalled, the aircraft

163

would experience a rapid bleed off of airspeed, a liability to spin, needing recovery techniques which were non-standard compared to conventional aircraft of the time and which resulted in a drastic loss of height. For these reasons stall warning was fitted and standard instructions were that if the aircraft was inadvertently spun and recovery had not been effected by 17,000 ft, then the aircraft should be abandoned. I only flew the Javelin T.3 and FAW.5 but in my limited experience the aircraft's controls were light at low speeds. It was very steady on an instrument approach but the control movements had to be fairly coarse to correct small deviations from the glide path.

At high altitudes the stick forces were light until above 0.9M when the elevator controls became heavier, however any manoeuvring resulted in speed being bled off fairly quickly. High throttle settings or loss of height were therefore necessary to maintain high Mach numbers. I would have liked a rate of roll at least double what was actually possible. At altitude the reduction in control response also meant that it was easier to experience pilot-induced over-corrections. Overall the Javelin was an advance on the RAF's previous night/all-weather fighters but it was overtaken by aircraft made by other companies (mainly foreign) who seemed to get much more performance from aircraft using engines rated at, or even below, those of the Javelin. It seems therefore that the problem was with the airframe design i.e. its weight, drag and size.

F/L Jim Sawyer spent just under four years on the Javelin having previously flown Venoms in the Middle East

I first entered the Javelin world in January 1959, on the 14th to be exact, when I flew my first sortie from RAF Leeming having joined 228 OCU for conversion to type and training in all aspects of air defence, day and night.

My first glimpse of the aircraft had me a little bit shocked. It was its sheer bulk and shape which made me wonder how on earth it could be a fighter being more akin to the bombers I had seen in my youth! Nevertheless, I was prepared to give her a chance.

My first sortie in the type, a Mk.5 light on internal fuel only, went better than I had expected. It had powered controls to which it seemed readily responsive and the pair of Sapphire engines gave it a kick better than that of a Venom or Meteor. Its instantaneous turning capability was good but there was a rapid associated reduction in speed. Thus it would probably throw off another fighter for the first 90–180 degrees of turn but after that it would have to go nose low and disengage. Early experience of landing the aircraft could be entertaining. While the speed of such was respectably slow it had 'British' toe brakes which were the first of its kind I believe. They were poorly graduated and as one applied them, were inclined to snatch. The consequence was often a weaving landing roll out until experience was gained.

Overall though, the Mk.5 had a respectable perform-ance compared with other fighters of the period and a higher operational ceiling than many. Like all its con-temporaries it had guns only although they were the formidable 30 mm cannon. It did however mean that to be fully effective, firing range needed to be less than about 600 yards. The Mark 7 et seq were very much better (as were the brakes) and they carried the wiring etc for air-to-air missiles. The gun only Mk.7 could give the Hunter 6 and others of similar ilk a good run for their money and often did. It had the more powerful engines and had an excellent rate of climb, acceleration and operational ceiling. Intercepts at above 45,000 ft were very feasible and became more so when the missile modifications were added.

My experience with the Javelin Mark 9 and 9R was limited to the last year or so of my time in the force and my impressions of it were generally very favourable. It had a simple reheat system which was negative in effect at low levels but became increasingly effective as height was increased above about 5–10,000 ft depending on ambient conditions. At low level it would underperform in comparison to the Mk.7 while remaining, of course, much more formidable than the earlier pre-7 models. At around 20,000 ft the power boosting came into its own and the aircraft was at its best and would outclass any other Javelin model as well as the Hunter and, indeed, USAF F-100s, although the latter could usually escape by virtue of its supersonic performance. That action however required the use of afterburner and the Firestreak air-to-air missile being a heat sensor meant that the escape had to be carefully planned and timed! It was usually via a head-on pass with little displacement where the Javelin crew would be forced to turn to achieve missile acquisition. By that time the opponent was usually out of the missile parameters.

I had some experience of firing the Firestreak when involved with development trials at Aberporth Range. While I was on the staff at the Fighter Combat School, the Air Fighting Development Squadron which, like us, was also a part of the Central Fighter Establishment sometimes needed to call on us to fill places when they were short of crews or were flying several other concurrent trials. Our targets were the Jindivik drones which could be enhanced as necessary to give them the characteristics of typical operational types, both friend and potential foe. In short, my experiences of trials with the Firestreak were encouraging and gave me confidence that it would be effective in combat.

To conclude, my overall impression of the Javelin Mk.9 was that it was a very positive improvement on all its forebears and an excellent interim aircraft before the Lightning. Its big advantage over the latter was that it had double the firepower, better range and endurance

166

and I believe was a better 'bomber destroyer' until such times as the Soviet Backfire came into operation. The aircraft handled well and could in my opinion be an effective fighter v fighter platform, particularly where the opponent was a single-seat aircraft. In various engagements I undertook on training sorties the benefit of two sets of eyes, within a well co-ordinated crew, were very considerable. The Javelin was an aircraft with which I was very happy to be associated and incidentally, it could be looped in safety despite a folklore which suggested otherwise!

Jim Sawyer continued to fly Javelins until September 1962 by which time he had accumulated a total of 450 hours flying time on type. He was then posted to a US Marine Corps squadron to fly the F-4B Phantom and later flew the Lightning with the Air Fighting Development Squadron and 23 Squadron.

Although the Javelin never fired its guns in anger there was more than one occasion when it came close and F/L Hank Prosser of 41 Squadron was involved on one such occasion as he relates in the following account

The squadron had gone to Geilenkirchen in holiday mood as a temporary home while our base at Wattisham was in the hands of the civil engineers. However not long after we were embroiled in the Russian attempt to deny the Allies access to Berlin via the air corridors. President Khrushchev was banging his shoe on UN tables. Instead of returning home to the warm embrace of UK pubs and our favourite bitter, we were planned to escort BEA Viscounts flown by RAF crews (no fare paying passengers) down the central Berlin air corridor with similar operations by French and American forces either side of us. Apparently the fact that we carried air-to-air missiles was supposed to strike terror in the hearts of our foes. The holiday mood was instantly replaced by something less pleasant as we were briefed on the forces ranged against us. Eastern Germany seemed to be a nettlebed of surface-to-air missiles, and in the air we appeared to be a little outnumbered, about ten to one as I recall. We had barely enough fuel to get there and back, straight and level.

We were quite unimpressed by the remarks of the local station commander briefing us on this as he concluded by saying 'Navs, take your cameras, you should get some interesting footage!' Several times a day we had conflicting orders from NATO, 2nd Tactical Air Force, Fighter Command, SHAPE, possibly even the Vatican as we were deployed between Geilenkirchen and Gutersloh and back again. Having arrived in Germany for a couple of weeks jolly, many of us felt that we hadn't even said a proper goodbye to those we had left behind with a distinct possibility that we never would. Throughout this sombre period our squadron commander, W/C Dave Smith, was brilliant and gave one of the most impressive displays of leadership I have encountered. Mercifully this operation was never put into actual effect, although the preparation remained in place for quite a long period with Javelin squadrons, and later Lightning squadrons, rotating the standby duty.

CHAPTER 16

Crew Debrief – 2

Having flown the Hunter F.1 and F.4 at Jever in RAF Germany with 98 Squadron, F/L Gerry Shipley later flew Javelins with both 23 and 60 Squadrons. Not long after commencing his spell on Javelins he was offered the chance to fly one to the Far East as part of a ferry operation

In the Spring of 1961, soon after my navigator, Bill Meads, and I arrived on 23 Squadron our Flight commander asked us whether we would like a trip to Singapore in the summer. We agreed that this sounded interesting and accepted the offer. It turned out to be a more interesting trip than we had expected.

In July we were one of four 23 Squadron crews sent to Waterbeach to join the 60 Squadron Ferry Flight Detachment. We joined one 64 Squadron crew and one former Ferry Wing crew, F/L Ted Owens and Master Navigator Tony Melton. In May/June, Ted had led a 'convoy' of four Javelins to Singapore and now our task was to deliver another six aircraft. Our convoy would be accompanied by a Hastings carrying our support ground crew and essential spares.

The planned route was Waterbeach – Istres – Luqa – El Adem – Diyarbakir – Teheran – Bahrain – Sharjah – Masirah – Karachi – New Delhi – Benares – Calcutta – Rangoon – Bangkok – Butterworth – Tengah. The aircraft were standard FAW.9s as the under-wing drop tank modification to the Javelin (FAW.9R) had not yet appeared. The only navigation equipment fitted was a radio compass and a Rebecca DME; there were no Eureka ground beacons for the DME between Nicosia and Butterworth. The entry on the Flight Authorisation Sheet was neat and to the point : 24 July, XH846 Shipley – Meads, Waterbeach to Tengah, as briefed.

One aircraft needed a starter motor change in Malta but when the pilot attempted a ground run after the replacement, the starter exploded through the rear cockpit (fortunately empty) to exit the far side of the fuselage, resulting in Cat.5 damage. Our convoy was therefore reduced to five Javelins, but at least we now had a spare crew in the Hastings. The next eight legs went smoothly, but crossing the Indian sub-continent did not.

During departure from Karachi, two of us were surprised to find the (ungauged) ventral tanks emptied on the climb out; the local refuellers hadn't mentioned that when they put in the quantity we had advised, much of it had exited through the pressure relief valves. We reached Delhi only by means of using the Javelin's maximum range descent (one engine shut down and the other at idle which gave us 120 nm from 40,000 ft for very little fuel cost. In New Delhi we found the Ashoka Hotel was particularly comfortable; Ted decided we all had crew fatigue and we stayed for two nights to recuperate.

Bill and I also had an unscheduled night stop in Benares having lost a generator. Ted Owens therefore took three aircraft on to Calcutta, leaving two of us to follow the next day. Overnight Bill complained of

170

an abdominal upset and it seemed sensible that he should fly to Calcutta in the Hastings so having a spare navigator thus paid dividends [Meads had to have his appendix removed shortly after arriving at Changi].

The weather for the next two legs was very poor; we had caught up with the Inter-Tropical Convergence Zone. The flight from Benares to Calcutta was a very uncomfortable seventy minutes of close formation flying, mostly in cumulonimbus clouds. On 5 August the weather for the flight to Rangoon was again very poor. This time I was flying with 'Benny' Baranowski as No. 2 in the leading vic of three (Ken Pye and Alex Saunders were No. 3) and once again we were caught up in monsoon-style CuNims. As we climbed through about 30,000 ft there was a bright purple flash and metal fragments were ejected from the upper fuselage of Ted's aircraft which immediately pitched nose down so fast that Ken Pye and I had no time to react. We had simply lost our leader in thick cloud and there was no response to our R/T calls. At Rangoon we had confirmation that Ted's Javelin had crashed, but it wasn't until we reached Tengah that we heard that Tony Melton had survived his ejection and been rescued after spending three nights in the Ganges delta. Ted has also ejected but had not survived. Having departed the UK with six aircraft, we therefore arrived at Tengah on 8 August as a box of four.

A Court of Inquiry was convened immediately with an engineering branch Wing Commander in the chair. When I described what I had seen of the failure of Ted Owens' aircraft he asked me what I thought had happened. I was aware that in the earlier days of Javelin operations there had been a number of catastrophic failures of the Sapphire turbines and said that I believed his starboard engine turbine had exploded. The Wing Commander produced a model Javelin and asked me to indicate whereabouts the event had occurred. When

I did so he remarked that this was well forward of the turbine and was closer to the compressor. When I suggested that in that case the compressor had exploded, he dismissed this saying 'Compressors don't explode'. It took some months and more incidents before the phenomenon of centre-line closure of the Sapphire's compressor was diagnosed but I am quite sure that I was the first person to see one happen. [The Javelin that crashed was XH791 and post-crash investigation by Martin Baker also found a good deal of evidence to suggest involuntary ejection. There was no emergency call or warning to the navigator, he merely heard a change in engine note, following by a bang and realised that the pilot had ejected. The pilot's body was found the next day hanging in trees with his ejection seat still attached. Damage to the seat was inconclusive however as he was wearing gloves, but with another glove clenched in one hand, it was considered unlikely that he would have voluntarily fired his seat while holding a spare glove.]

The high cost of the Firestreak missile system precluded their widespread use for practice purposes. During his time on Javelins Gerry Shipley was one of the lucky few who were fortunate enough to be given the chance to fire a Firestreak as he recalls

We were tasked with a couple of 'extras' in addition to launching (I believe) five Firestreaks during 23 Squadron's detachment to Valley in May 1963. One was to enable Mike Chase, then the undisputed doyen of air-to-air photography, to get a picture of a Javelin having just launched a Firestreak, the photo to include the aircraft and the missile. The other was to enable C-in-C Fighter Command to watch a firing. I was designated pilot for the second firing and to fly Mike Chase to photograph

the first one, which C-in-C would observe. Range qualification required the usual practice sorties and was
completed in a couple of days, but the programme was
then put on hold. It appeared Llanbedr were concerned
about the loss rate of Jindiviks; although the Firestreaks
were all telemetry rounds, without warheads, there had
been an unacceptable number of direct hits. The drones
were being adapted to tow magnesium flares which
were intended to seduce the Firestreak away from the
Viper jet pipe, but so far no modifications had been
completed.

After about a week someone's patience ran out and
it was decided to allow a firing using an unmodified
Jindivik, hoping that it wouldn't achieve a direct hit.
The sortie would also include the C-in-C in a Javelin
T.3, to afford the best view, and me flying Mike Chase
for the photograph. It all worked well, C-in-C was
delighted and Mike's photo was very impressive and
featured on the front cover of Fighter Command's
'Flight Safety' magazine a month or two later. The
only dissatisfied participants were Llanbedr who were
miffed at losing yet another Jindivik. It was decided
that no more firings would be programmed until a flare-
equipped Jindivik was available. To fill in the week
of inactivity while this work was completed, we were
treated to an Anson flight to Aberporth to admire their
radars and party in the Mess, followed by a bus trip up
the coast to Llanbedr to admire their Jindiviks and
party in their Mess.

It was therefore over two weeks after our arrival at
Valley that my range sortie was programmed. It was to
be a routine launch at a target at 40,000 ft. The weather
was perfect and the only minor hitches were crew
induced. My usual navigator had gone down with a
cold and Sam Bass who filled in was less well rehearsed
and I suspect slightly hung over, which led to a less
than perfect example of dialogue on the aircraft wire

recorder for the debrief. To enable Aberporth to start their telemetry recording at precisely the right moment, the navigator was briefed to transmit a steady countdown '5 4 3 2 1 FIRE', so that the buttons could be pressed at the appropriate stage. Sam forgot, trod on the R/T pedal and called '1 2 3 4 5 FIRE', consequently the high speed recorders at Aberporth didn't quite get what was wanted, since there was something of a rush to the switches when they heard the number '1'.

The aircraft wire recorder then carried on with:

Self (Intercom) – 'Missile is away, veered off to the right, correcting, homing, it's hit the flare, now it's hit the Jindivik'

Self (R/T) – 'Aberporth, Eagle Red 1, exercise complete, OFFSIDE'

Navigator (Intercom) – 'What happened Gerry, did it launch OK?'

Self – 'Weren't you listening Sam, we just shot the Jindivik down'

Navigator – 'F###### H###'

Self – 'Hey Sam, have you turned the wire recorder off ye ......'

However, it was an excellent day's entertainment; the Firestreak performance was very gratifying, the only people not too impressed were the staff at Llanbedr who were unhappy that even the flare modification hadn't saved their Jindivik. I can still remember clearly the sight of the little red aircraft dissolving into an orange ball, surrounded by black smoke against a clear blue sky. I hadn't been prepared for that, the MPC films we'd been shown were all in black and white!

A year before the story related above, Gerry Shipley had what might be termed an 'I Learned About Flying (Javelins) From That' moment as a result of trying to keep up with a much faster aircraft at low level

A flight of Royal Swedish Air Force Saab Drakens were to visit Coltishall for a week and it was decided that it would be a hospitable gesture to intercept then inbound and escort them to the airfield. The OC 'A' Flight and I were to provide a pair of 23 Squadron Javelins and 74 Squadron would provide a Lightning. We met up with the visitors at about FL 350 and so had a combine of four Drakens, one Lightning and two Javelins. As the Drakens had UHF radios and we were still fitted with VHF, communication was difficult, but GCI relayed to us that their leader intended to do a flypast on arrival and had invited us to join in.

The approach to Coltishall was made in a continuous descent and the Draken leader clearly had sporting views regarding flypast heights; as we approached the airfield we were certainly not much above hangar roof level. At this stage I noticed that OC 'A' Flight was no longer with us and my aircraft was much noisier than usual. All of my attention was focused on the Draken on which I was formatting, so I asked my navigator for an airspeed check. In the backseat Jim was fully occupied with his camera, but took a quick look at his ASI and told me '540 knots'. The Javelin was limited to 535 kts IAS, but I decided that five knots over the top was acceptable and stayed with the fly-by.

After the flypast, decelerating to leave the formation proved difficult; throttling back produced some strange noises and the airbrakes seemed not to have much effect. Swapping speed for height worked, but the height gain was more than I had expected. It seemed that in his quick glance at his multi-needle ASI Jim had misread, we had in fact reached 640 kts IAS. The OC of 'A' Flight had recognised what was going on and had sensibly pulled out. He had tried to tell me to do likewise, but within the noise being generated by our own aircraft, Jim and I had not been able to hear the R/T. In post-flight conversation with my own Flight commander, Mike 'Dusty' Miller, I said how surprised I was that the Javelin was capable of such a high IAS. He then gently explained to me that the limitation had been imposed, not for any shortcoming in performance, but because of the risk of flutter which had in the past led to the elevators becoming detached. He was also kind enough to decide that the incident was just part of the learning curve!

After flying Javelins, Gerry Shipley was an instructor on Gnats at 4 FTS Valley before joining 43 Squadron at Leuchars where he flew the Phantom FG.1. On leaving the RAF in 1974 he was twin-engine flight manager at Air Service Training at Perth until 1987 when he became CFI at the British Aerospace Flying College at Prestwick, a position he held until 1996.

Like a number of other RAF pilots in the mid 1950s, F/L Fred Butcher made the transition from single-seat day fighters to two-seat night-fighters. Any reservations as to whether he had made the best of career moves were quickly allayed when he came to fly the Javelin

Having trained on the Meteor, it seemed like a downward step to be posted to 118 Squadron in RAF Germany and on to Vampires. We then got the Venom which was a great improvement, followed very quickly by the Hunter, which was vastly superior. When posted back to England in 1955 it was another downward step to the Venom NF.3 on 23 Squadron. However we knew that the Javelin was coming and when it arrived in May 1957 it was a giant leap forward when compared with the Venom. Like the Hunter before, there was no dual control version, so it was 'Get in and fly it'. I loved it from the word 'Go' and it was a real pilot's aircraft. The air conditioning was first rate and one was comfortable at all levels, unlike other aircraft I had flown. It was very responsive on the controls and despite its size and weight, it was very manoeuvrable. Later, I was chosen to give a display at the Coltishall Open Day, and later still when on 60 Squadron in the Far East, I gave aerobatic displays at Tengah, Butterworth and Paya Lebar.

One of the Javelin's drawbacks was the fact that it would not recover in the conventional way from a stall. This was due to the delta configuration and high-set tailplane. Supposedly you had to go into a spin which took 15,000 ft to recover. For this reason aerobatics in the looping plane were prohibited. I flew the aircraft over a period of ten years and never had a problem with this, although I know that there were a few crashes resulting from stalls. I believe the Javelin would have been a much better aircraft had it been equipped with Rolls-Royce engines as many accidents and incidents were caused by the Sapphires that were fitted. Centre-line closure caused quite a few mysterious and sometimes catastrophic failures until the engineers found out what was causing it and modified the engines. The AVPIN starting system also left a lot to be desired and there were many starting failures, some resulting in the loss of the aircraft.

The hydraulic system on the Javelin was a good one with four pumps on the earlier aircraft and five on the later ones. There was no manual control reversion with loss of hydraulic pressure in the right (or perhaps wrong) combination of pump failure, so this would mean abandoning the aircraft. Of course, in the event of a double engine failure it would also be ejection time. I had personal experience of this but luck was definitely on my side. I was about four miles out on a night approach to Coltishall when my starboard engine ran down. I tried an instant relight which was unsuccessful. I wasn't worried as the performance on one engine was excellent, especially at landing weight. Also I was only at 500 ft and about a mile from landing. The landing was normal, but as I was about to clear the runway the other engine ran down. I had had a double engine failure, all within about two minutes. Had the second engine failed two minutes earlier it would have meant a double ejection on the approach. In all my ten years of flying the Javelin I had many minor failures of no real consequence, but this was the only major failure and I didn't really know about it until it was all over!

Later in his career on Javelins Fred Butcher flew with 60 Squadron in the Far East where he witnessed at close quarters another aircraft suffer a catastrophic structural failure

On 60 Squadron the first programmed take off consisted of two pairs with a designated leader. The four aircraft would split into twos who would then practice high-level PI's on each other. Then towards the end of the sortie they would join up into a four for a tail chase to be followed by a run and break. On the morning of 10 February 1964 there was a delay waiting for a fourth aircraft which did not become serviceable. I was designated leader so I briefed that we would do the PI's on each other and then form up for a three-ship chase. Since I thought that I needed some practice in following, I said that Bob Lockhart would lead the tail chase then Gordon Sykes/Paul Burns as No. 2 with me and Paul Rundell as No. 3.

It was a beautiful sunny Singapore morning as Bob commenced the tail chase. As I recall it was all going well, very smooth with no high 'g' forces involved. I had kept about 300 yards station behind the No. 2. We were at about 18,000 ft and pulling about 3g after a barrel roll when I saw what looked like a lot of tinfoil coming from the base of Gordon's fin. Before I even had chance to shout out, the whole of the fin and tailplane broke off, narrowly missing us as it went underneath.

There was no transmission from the No. 2 so I made a Mayday relay call to the controlling ground station.

The stricken Javelin was miraculously still straight and level so I flew alongside in formation. Both aircrew ejected at the same time and I followed the parachutes down till I saw them land in the sea. I also watched the Javelin in its falling-leaf-like descent till it crashed into the sea. I then flew over the two dinghies and could see that one was in his dinghy and the other was clinging to the side. I knew that help was on its way, and in any case by this time I was getting low on fuel so returned to Tengah.

The navigator of Javelin FAW.9 XH747 was F/L Paul Burns who recalls the accident from his perspective

The weather for the sortie over the Malacca Straits was perfect with no cloud or turbulence and with little for me to do except maintain radar lock. The sortie was proceeding without a hitch and I took the opportunity to enjoy the early morning record show on Radio Singapore. In due course we were flying as No. 2 being led by Bob Lockhart's aircraft and trailed by Fred Butcher in a moderate tail chase at 18,000 ft. Suddenly with no warning, there was a loud noise and, as I remember, a severe rolling sensation. It was immediately obvious to me that the aircraft was no longer under control of the pilot who was in extreme stress and agitation. Concurrently I heard Fred call 'Mayday, Mayday, Mayday, Javelin aircraft breaking up.' Without further thought or hesitation I took the decision to eject. Time stood still and it seemed nothing was happening and I was turning to look at the blind to see what the problem was when the canopy went and was immediately followed by the seat and myself. I know that Bill followed closely behind having used, in extremis, his seat handle [Gordon Sykes had the nickname 'Bill' on 60 Squadron after the character from Oliver].

I remember little of the ejection but recall the automatic separation from the seat which was at the preset height of 10,000 ft. The parachute deployed as it should but in the process I became aware of considerable pain in my lower back. Everything was quiet and still and the view was amazing. A short distance away was a large three-masted junk in full sail, moving away from the area. Throughout my descent it continued to sail away. In the descent I remembered my survival training and attempted to release the dinghy attachment clips but was unable to achieve this because of the pain. Consequently I hit the water before I could inflate the dinghy, the parachute remaining attached, thus acting as a huge sea anchor. Bill landed in the sea some 200 yards away but was unable to get into his dinghy which also had the 'chute attached. He was conscious but in great distress. Our position was also made worse by a Javelin circling overhead at very low level causing huge waves by its jetwash [Fred Butcher had already left the scene at this point as he was low on fuel]. After about half an hour the rescue helicopter arrived.

The helicopter went first to recover Bill who was still in distress in the water clinging to the dinghy. On dropping the winch ahead of the linesman the swinging ball struck Bill a severe blow on the head. Then on his arrival the crewman, having put the harness around Bill, gave the sign to lift but this proved impossible because of the parachute acting as an anchor. He was dropped back in the water, probably causing more injury, and was eventually lifted and secured in the helicopter which then lifted me. Once on board I could see that Bill was barely conscious. Within thirty minutes we were landed at Changi and taken to hospital where we shared a ward. Within three days it was confirmed that Bill had badly injured his spine and on the

Thursday he was casevaced to Brize Norton and on to Stoke Mandeville hospital. He was paraplegic for the remainder of his life where he rose to be a senior staff member with British Motors. On repatriation to the UK in May 1964 I spent three months at Headley Court before regaining my flying category. I was then posted to 64 Squadron at Binbrook, a Javelin squadron soon to be deployed to Tengah during Confrontation.

The accident caused much disquiet, not only at Tengah, but throughout the upper echelons of Far East Air Force and Fighter Command. The ramifications were such that some officers in positions of authority would have preferred the whole matter to have been quietly forgotten about as Fred Butcher recalls

Although it happened forty-seven years ago I remember it as though it was yesterday. The powers that be were very reluctant to ground the Javelin. So much so, they tried to suggest to me that I hadn't seen what I said I saw. However I stuck to my guns and the Javelin was grounded for a couple of days while all the aircraft were checked. The engineering opinion was that it couldn't possibly have happened, but both Paul Rundell and I knew that it did.

I have my own theory as to why if did happen which came to me later after a minor incident with another aircraft. Because the Javelin had power controls, artificial feel had to be built in. This was achieved by having two artificial feel simulators in the hydraulic system. These worked very well and you could not easily pull too much 'g'. Some months after the accident on a routine check, one feel simulator on an aircraft was found to be unserviceable. The aircraft had been flown for some time like this and no pilot had noticed because one feel simulator was as effective as two. Now what if one had failed in XH747 with no-one noticing and then the other had failed in flight? With no feel it would be possible to deflect the tailplane fully and at those speeds it would have overstressed the tailplane. Of course, there is no way of knowing for sure, but I believe that this is what happened.

Paddy Harbison came to the Javelin having flown Spitfires and Mustangs during the Second World War and F-86 Sabres with the USAF in Korea. He took over the command of 29 Squadron in January 1958 as it swapped Meteor NF.11s for the Javelin FAW.6

The reason I was given command of 29 Squadron was that up to that time the night-fighter world had been very close-knit and generally the crews were pretty old. After the CO of 46 Squadron killed himself in a Javelin, a regulation was brought out that you couldn't be over thirty-four to take command of a Javelin squadron so they had to go into the day world to get people in that age bracket. I was serving in the MoD at the time when I was suddenly told about a new command. Of course most day fighter chaps were ready to land when the clouds began to lower and it got dark, so it was a bit of a shock to the system to go off on the first Javelin night-fighter course at Leeming.

The Javelin was a very large aeroplane and weighed almost as much as an F-4 Phantom. It was quite popular but I would say that it gave you a false sense of security when you looked out at that big wing. Its stall characteristics were not good for the role that it was in due to the fact that the wing blanked off the tail at high angles of attack. In general it handled well as long as you remembered it was a bit dodgy when pulling 'g' at low speed. The FAW.6 had the American-designed radar which had a scanner that allowed us to search when we had locked on, so we could still see what was around

182

us. Once you had locked on with AI.17, which was the British radar, you were blind to anything else

Occasionally we used to intercept B-36s coming across to the UK and for us they were very slow and usually at too low an altitude for comfort. As you slowed down to come in from behind, up came the nose and you were always very worried. The navigator would sometimes give you impossible directions as he was only looking at blip displacements. He would often tell you to go hard left or hard right but at low altitude we daren't go hard left or hard right as we were very close to the stall anyway. What we used to do was to press the 'Test' button so that all the stall warners would jangle and this was guaranteed to get the attention of the navigator who would quickly say 'Straighten up!'

Paddy Harbison was to retire from the RAF in 1977 with the rank of Air Vice-Marshal having spent the last two years of his RAF service as AOC No. 11 Group with responsibility for the Air Defence of the UK. Having already been awarded an AFC in 1956, he was later made CB, CBE. Before that however he was station commander at Leuchars from 1963–65 which was where F/L Pete Thorn encountered him during his service as a Javelin pilot with 23 Squadron

Paddy was a well-liked Boss and had a tremendous sense of humour. I remember one Friday night beer call in the Mess (a standard habit at the end of a working week in the RAF) when he came up to me at the bar and asked me how many beers I had had. When I told him that I was starting my second he said 'Right, put it down, go and get your flying kit on. I want you to fly me to Ireland, my son has fallen and broken his leg. I will fly it there, drop me off and you bring the aircraft back again.' As we got to height and contacted Irish Radar they came back on the R/T and asked what aids we had (it was a Meteor T.7 with no aids but the

JAVELIN FROM THE COCKPIT

radio). Quick as a flash Paddy replied 'I'm talking to you aren't I . . .?'

On another occasion when flying a Javelin I had been diverted because Leuchars had gone down in fog. This was not unusual at Leuchars (it was known as the 'Haar') as it was very subject to fog coming in with the tide, especially in Autumn. I arrived back the following day just as the fog was coming in again and it was already on the threshold of the airfield. Having broken out of the approach I overshot, and at low level with undercarriage and flaps down, positioned myself to land from the opposite direction. I was low and took out the married quarters for a tight circuit. By the time I had landed, parked and shut down Paddy arrived at the foot of the access ladder and gave me a Paddy 'telling off', but in his usual way he didn't make it so – 'Peter, I'm glad to see you back; it was a good job I had my front and backdoor open; would you kindly make your overshoots a bit higher next time, OK?'

Pete Thorn had joined Avro as an apprentice at the age of sixteen (his father was S.A. 'Bill' Thorn who was the company Chief Test Pilot). He was then accepted for the Fleet Air Arm but following cutbacks in the demand for pilots he returned to Avro as a flight test observer where he flew in aircraft such as the Lancaster, Lincoln, Tudor and Athena. He then joined the RAF and was to spend several years flying Meteor night-fighters with 25 Squadron at West Malling and Tangmere. Having left the service for a time, he re-joined in 1962 and was posted to 23 Squadron to fly Javelins, initially at Coltishall

I served in the RAF in the night/all-weather world flying Meteor NF.12 and 14's. That was until Duncan Sandys, the politician, ruined a lot of our careers by virtually disbanding all of Fighter Command from 1954 to 1958. I was one of the many who resigned from the

Air Force as we were not interested in becoming Air Traffic Controllers or Radar Operators. In 1962 I rejoined as they realised that Eastern-facing missiles around the East coast were not as useful as Fighter Command. Going back into the RAF was no problem and I was posted to 23 Squadron at Coltishall, moving to Leuchars the following year.

No. 23 Squadron was selected as the in-flight refuelling squadron and we did a lot of training which included trips to Malta, El Adem, Aden and Cyprus. In-flight refuelling in calm conditions was reasonable, but in turbulent conditions life became interesting. The hose was reeled out quite a distance and when it was turbulent the hose and the fuel connection used to wander all over the place, such that we would often ask the Valiant tanker to execute a turn to put some 'g' on the hose which would tend to reduce its 'wandering' and enable us to engage with the connection probe. The probe on the Javelin was rather like a jousting tube, situated along the top of the fuselage to the right-hand side. As we were night/all-weather it could get quite exciting for those awaiting their turn at night, in turbulent weather.

In July 1963 it was decided that two aircraft would do an overnight detachment from Leuchars to Cyprus. I was lucky as my navigator, Dave Hodges, and I were selected to go with the CO as his No. 2 on this detachment which was the first ever night refuel. It happened to be one of those turbulent nights and the Boss had some difficulty engaging on our first refuel over Calais in France. When he made contact with the dish, he engaged so hard it put out the four small guide lights which were there to help you engage in the dark. The result was that I had to ask the tanker to illuminate the underside of the aircraft so that I could silhouette the dish against the fuselage. Other than that we had an enjoyable run with the sun coming up for the last hour of the flight. We landed at Nicosia after 5 hours 55 minutes.

Eventually the Valiant tanker fleet had to be scrapped as the aircraft had main spar problems. The next aircraft which we could refuel from was the American KC-135. This actually made our refuelling a lot easier. USAF aircraft refuel from a tube (or boom) which is extended and controlled by a member of the tanker crew who guides the probe into the aircraft. In our case with a fixed probe, they attached a short flexible tube with a 'saucer' on the end. This made refuelling, for us, a 100 per cent improvement. There was no more chasing a line with a mind of its own, whatever the weather. Of course, we had our own radar so we could pick up the tanker in cloud, even at night.

Overall I enjoyed flying the Javelin and it was certainly much more comfortable than the Meteor. The Javelin FAW.1 was a very different aircraft to the FAW.9 which had an all-moving tailplane, increased fuel tanks, improved radar and four Firestreak infra-red homing missiles, and more powerful Sapphire engines developing 12,300 lbs thrust in reheat. The early marks of Javelin were not very popular but with the advance in technology, pilots began to appreciate its excellent handling and endurance. She was supersonic (just) and also had very powerful airbrakes such that if you were going to apply them you warned your navigator in case he had some items loose (pencils etc) which would very quickly disappear. If any Hawker Hunter spotted us they would invariably try and get on our tail. This we let them do but as he came into our rear view mirror, you applied the airbrakes. The Hunter would then overshoot us very quickly, we would put the airbrakes in and as he appeared in front of us, take a photo and send it to the Hunter squadron with a caption saying the time and date and ... shot down by a Javelin!

The Javelin's reputation for successfully fulfilling operations far removed from its original 'bomber destroyer' role, was further enhanced in 1965 following Rhodesia's Unilateral Declaration of Independence. Apparent tensions with the newly independent Zambia, in particular over the jointly owned Kariba Dam which provided hydro-electric power for Zambia's copper mining industry, led to the deployment of 29 Squadron's Javelin FAW.9Rs for a ten-month period. Pete Thorn was one of those involved

As the Valiant had just been permanently grounded we were not able to in-flight refuel and so had to 'puddle jump'. We departed Akrotiri on 28 November 1965 and flew to the first overnight stop at Bahrain via Diyarbakir (Turkey) and Dezful Vahdati (Iraq). The following day we continued to Aden via Masirah (Oman) and on 1 December we flew to Nairobi in Kenya. There was then a delay whilst we got permission to continue and it was not until the 3rd that we arrived at Ndola in Zambia (formerly Northern Rhodesia). We were then grounded for ten days awaiting tankers to arrive by air with fuel for the aircraft, during which period we were visited by Kenneth Kaunda, the President of Zambia.

Having arrived, we were then stuck. The length of the Ndola runway was too short for us to fly back to Cyprus as full fuel was needed to get to our first refuelling point at Nairobi. Later on they lengthened the runway by 500 ft! As we got our feet under the table as it were, half of the squadron went down to Lusaka which was where I spent most of the ten-month detachment. Domestically it was a wrench as we were based in Cyprus which left the families trapped, unless you had a lot of cash (which we didn't have) for the girls to go back to the UK for a break. However, once organised and settled we had a rotation of crews back to Cyprus every three months for six weeks. During operations in Zambia reheat was used on take off as the airfield altitude at both Ndola and Lusaka was some 4,000 ft amsl.

Probably the most amazing thing after our arrival was the tremendous welcome and hospitality we had from the population. Doors were opened and it was a case of please come round, stay the night, use the swimming pool etc. All the sports organisations offered us honorary memberships and they were not just being polite; they really did mean it and were rather offended if you did not take up the invitation. We were also invited to take part in the Christmas celebrations, if we had an item. As it turned out we had, something known as the 'Whistling Navals'. To explain, we had a selected team of six of us stripped to the waist, with large made-up paper/card bowler hats into which we put our arms and head. The exposed waist was then decorated with painted eyes around our nipples and pursed lips around the navel. A jacket was put on a coat hanger which was hooked to the back of your trousers and buttoned up around your legs. To the whistled tune of Colonel Bogie, (we had a gramophone record which we always kept on the squadron) we pulled our tummies in and out to the tune. It brought the house down and we had to do three encores!

Having flown nearly 1,000 hours on the Javelin, Pete Thorn later became a QFI and served on the University of Wales UAS before being posted to Gutersloh in Germany to supervise the Lightning F.2A simulator (he was also Station Flight Safety Officer). His next tour was as a deputy Flight commander at the RAF College at Cranwell on Jet Provosts and after three years he joined the Flying Selection Squadron at Swinderby as an examiner, flying Chipmunks. During his time at Cranwell he also enjoyed a 'magic' 2½ years with the Battle of Britain Memorial Flight flying the Spitfire and Hurricane and this was to lead to the opportunity to fly several privately owned aircraft, including Spencer Flack's all-red Spitfire XIV G-FIRE. After leaving the service he became CFI for private pilot training at Oxford Airport.

After serving with 60 and 64 Squadrons, F/L Paul Burns later joined 29 Squadron and also took part in the deployment to Zambia. He describes some of his experiences during this time and another ejection when landing at Akrotiri in Cyprus

In December 1966 No. 29 Squadron had only recently returned to its base at RAF Akrotiri in Cyprus from a year-long operational deployment to Zambia to provide an air defence capability against the perceived threat to the integrity of its airspace posed by Southern Rhodesia following UDI and, in particular, to provide air defence for the recently completed Kariba Dam. The aircraft were evenly deployed at the two Zambian airfields able to operate jet fighters, Ndola in the Copper Belt Region and closer to the border at Lusaka Airport where a dawn-to-dusk readiness state was maintained. A secondary purpose was to deny airfield space to the USSR.

Ndola airfield was over 4,000 ft elevation and the runway was just over 2,000 yards long. It was made of asphalt and thus did not allow the standard procedure of aligning for take off and running up the engines to 70 per cent power on the brakes before release because we blew chunks out of the runway. With massively reduced aircraft performance we only just got airborne and started slowly to climb away. By contrast BOAC jets used half the runway and were soon in a steep climb.

To provide sufficient Javelin aircrew for this demanding task, the recently closed Javelin Operational Conversion Unit in the UK had to be re-opened in haste to provide the additional aircrew needed by the squadron to allow crews rest and recuperation in Cyprus on a man-for-man replacement basis. This ad-hoc roulement

189

combined with little opportunity for the crews in Zambia to get airborne because of the fuel embargo imposed by Rhodesia, resulted in total disruption of normal squadron organisation and, to a large extent, morale. The practice of having set crews was also not possible. The deployment was ended in September 1966 but squadron organisation and cohesion was only slowly restored.

On 14 December 1966 I was detailed to fly [in Javelin FAW.9R XH848 'L'] with F/O John Pierce, a recent graduate from Cranwell who had completed the course at the newly opened Javelin OCU. I believe he had less than 1,000 flying hours in total with very few hours on type. I had never flown with him before, but this was now the norm. We were programmed for take off at 0700 hrs as No. 2 in a four-ship formation authorised for high level 2 v 2 combat followed by rapid recovery in 'day fighter' combat formation for a high-speed break and stream landing on runway 09 at Akrotiri. The formation was to be led by S/L Tug Wilson, a highly respected and very experienced Javelin pilot with several thousand hours. The No. 3 was Master Pilot Tam Kirk, another highly experienced day and night-fighter pilot and No. 4 was F/L Bob Lockhart, a multi-tour Javelin pilot with very many hours on type. In this company John was a 'man among giants' and was under a considerable amount of pressure not to let them down.

It was a beautiful clear and cloudless early morning for the sortie and despite minor aircraft problems, the formation was airborne only slightly behind schedule. We were soon at the briefed height engaged in 2 v 2 combat. This was an unusual sortie for the Javelin but it went well and was enjoyable. In due course we joined up in a fluid-four combat formation and carried out a rapid descent and a high-speed break for landing at Akrotiri. The break appeared to be well executed and the aircraft were well spaced. However, known only to him, John as No. 2 had achieved his correct spacing by reducing aircraft engine power and speed and, with hindsight, should have left the formation and gone round again. To do this he would have lost all face and no doubt this led him to continue in the formation landing still keeping his correct spacing from the leader by further reducing power and thus putting the aircraft into a nose-up and low-speed turn onto final approach. We continued the approach but as the aircraft came on to the runway alignment we experienced jet-wash from the leader, causing John to apply full power and in doing so putting the nose-up aircraft into a fully developed stall. Without a moments hesitation I ejected knowing that the aircraft was lost. John closely followed. The aircraft crashed into the technical area a hundred yards or so from the threshold and was completely wrecked, but did not burn.

On a low-speed stall the Javelin had an extreme pitching motion and I had ejected with the aircraft nose-high, making my trajectory approximately level with the ground. Thus when the bolt which extracted the drag chute fired, it snagged around the seat and the main chute failed to deploy. The ejection height was judged to have been about 300 feet. I hit the ground still strapped in the seat but unconscious. The ground impact was with a smallish thorn brush and this undoubtedly helped. John ejected as the aircraft was more level and his seat functioned correctly with his parachute fully deploying. However, with insufficient height to allow him to swing to the vertical he hit the ground with force and broke a leg.

It was the usual practice at Akrotiri to have a rescue helicopter with engines running and at immediate call whenever flying was taking place. Thus it was only a matter of a few minutes until the rescue crew arrived on the scene. The winch man immediately assumed I was dead and reported so. However I then showed sign of life and without further ado he extracted me from the seat and flew me to the Military Hospital located on the other side of the airfield. The time from crash to admission into critical care was less then six minutes. This swift action undoubtedly saved my life and both crew members were awarded commendations for their rescue.

I had sustained eighteen fractures – three in the left arm, nine broken ribs, three fractures in the shoulder, three fractures of the pelvis, and there was also damage to my left ear. These would all heal in time. The most serious threat to my life was the very real possibility of my lungs flooding during the ten days I was unconscious under sedation. A team of two Chief Technicians, medical gymnasts by trade, visited me every four hours of that period to beat my lungs clear and I survived. Both were awarded a well-deserved Queen's Commendation. I was bed-bound on a stretcher device unable to move for a couple of months. Then I learnt to walk again. I was transferred by air to RAF Hospital Wroughton at Easter 1967 and to Headley Court in June. In November 1967 I returned to flying with a reduced flying category for Navigation Refresher Training and subsequently finished my flying career as Bombing Leader on 45 Squadron at RAF Tengah in Singapore.

After his first tour with 46 Squadron at Odiham, Colin Edwards later flew with 60 Squadron as part of the Javelin force in the Far East Air Force. In the following account he describes a very different type of service life to what he had become accustomed to back in the UK

The climate posed different challenges, fog or low cloud was no longer a problem but these conditions were replaced by regular tropical storms. These were usually short lived and could often be avoided when landing. I remember one occasion when the squadron was detached to Butterworth during the confrontation period. The weather was doubtful in the early evening (we usually flew the first pair just before 7 pm) but we decided to fly each night. Meanwhile an Australian Sabre squadron desperate to night qualify cancelled every night but decided to fly on the Friday when we chose not to (it was like a game of poker seeing who would chicken out). That night I sat on standby at the end of the airfield and watched six Sabres take off. Then a storm blew up and the Sabres returned and decided to land with the storm half way down the runway. They landed too far up and could not stop on the flooded surface. They waltzed down the runway, madly trying to avoid each other. Two went off the runway, two went into the barrier and the other two collided. Within five minutes the weather was clear, it was a beautiful evening and we watched them pull the damaged aircraft out of the mud. Very enjoyable!

When airborne the cumulonimbus clouds, which could go up to 60,000 ft, were to be avoided. When they couldn't be avoided we ran into problems with centre-line closure and our squadron commander (W/C Peter Smith) and his navigator had to eject after engine failure. The Boss was quickly picked up from the jungle, but his navigator spent several days waiting to be rescued. He was S/L Frank Joliffe, a rather suave suede-booted type who certainly wouldn't have paid much attention during out intensive jungle survival course. Days later when he was found I asked him whether he had done all the vital things we had been taught. 'No', he said, 'I just sat down and waited. I knew some little brown men would come along and rescue me.' This happened. Suddenly a little brown face popped out of the jungle and said 'Squadron Leader Joliffe, would you like a cup of tea?' Actually I stood to lose money because I was running a book on whether he would be found [this accident involved FAW.9 XH836 which crashed near Mersing in Malaya and occurred on 3 December 1962].

Squadron life in the Far East was very different to what we had been used to. All of us had been first-time crews on Javelins on our UK squadrons but now we arrived on 60 Squadron as very experienced crews from the pick of Fighter Command. It really was like a meeting of the top gun fighters in the old west. The old 60 Squadron Meteor crews knew a lot about how to fly in the tropics but nothing about how to operate Javelins. This led to amusing situations. Our long suffering deputy Flight commander had to put up with our constant suggestions of how the flying programme should be organised. This reorganisation was entirely based on improving our social lives. For instance we usually flew five pairs. The old way of doing it was for two pairs to fly followed by one pair which flew only once. The first two pairs turned round, had a cup of coffee and then went off again. We organised this so that the pair to fly once, flew first and then quit for the night. This meant that the other pairs had to rush their turn round, go without their coffee and take off without a break. The deputy Flight commander, Pete Gooding, thought this was madness and tried to veto it. We tried to argue on the grounds of operational efficiency but we really had no case. Finally we had to come clean. We had booked seats at the cinema at 8 o'clock. 'Christ', said Pete 'you even want time off on your nights on!'

When we were on days we flew only in the mornings and were in the bar at lunch time, slept during the afternoon and partied most evenings. When we were on nights we swam at the Singapore Swimming Club, or the Tanglin Club, every morning, played golf and then went in for met briefing around 5pm. It always got dark at the same time so we could get our flying over and be finished before midnight. Night flying suppers were often fun. If it rained there were always a lot of frogs around. These could be put to good use. Sometimes they would be placed under the overturned teacups besides everybody's place ready to jump out when lifted. At parties they could be slipped into the mugs of beer of unsuspecting drinkers. Often they would drink on impervious to the frog frantically swimming around.

The flying was very enjoyable. We did a lot of low level interceptions around the islands off the Malayan coast. My pilot reckoned that the confrontation was caused by him trying to blow over junks by flying fast and low over them and pulling up when over the top. We had a lot of fun fighting against Australian Canberras and Sabres and we were excited by the thought of fighting the Indonesians but a little puzzled as to how we would intercept a Mustang or a Dakota. I think a Dakota could have flown at 120 knots and just rolled a big bomb out of the back. After leaving the RAF I went into teaching and to illustrate how stressful teaching can be I would often say that I have been upside down at nought feet over the jungle without a care in the world, but I have nearly had apoplexy on playground duty!

CHAPTER 17

Javelin Swan-Song

With the introduction of the English Electric Lightning F.1 to squadron service in July 1960 it had been assumed that the Javelin would gradually be phased out in favour of the new single-seat interceptor. In view of this the Javelin Operational Conversion Unit (No. 228) was disbanded at Leeming on 15 September 1961 as it was felt that the training of new crews was unnecessary and that there were already sufficient pilots and navigators to fulfil the RAF's needs as the number of Javelin squadrons was pared back. In terms of UK air defence the Lightning did take over the interceptor role as intended, however the need for continued all-weather fighter defence in RAF Germany and the Near East until these duties were also taken on by the Lightning, together with the confrontation between Indonesia and Malaysia in the Far East, led to the Javelin's service life being extended by several years. As a result of the Javelin's unforeseen longevity the decision was taken to re-form 228 OCU and this unit came into being once more at Leuchars on 1 May 1965.

The choice of Leuchars as the new base for 228 OCU was down to a number of factors, one of which was the fact that the Scottish base could boast a Javelin FAW.9 simulator and it was also home to the Javelin T.3s of the Fighter Command Instrument Rating Squadron. The aircraft that were used by the OCU were the FCIRS Javelin T.3s and FAW.9s that had previously been flown by Nos. 5 and 11 Squadrons at Geilenkirchen in Germany, with others being delivered via 27 MU at Shawbury. Pilots new to the Javelin would initially fly the T.3 to solo standard before progressing to the FAW.9. By this time they would already have been allocated a navigator and the new crew would then work their way through ground school and the flying programme. The course comprised about 50–60 hours flying time for pilots with about 45 hours for navigators. In addition a similar number of hours would be spent in the simulator. Weapons training involved practice interceptions using Firestreak acquisition rounds.

The new crews that were trained at Leuchars allowed Nos. 60 and 64 Squadrons to continue their vital work policing the skies over Malaysia. No. 60 Squadron had long been associated with the Far East and had swapped its Meteor NF.14s for Javelin FAW.9s in July 1961. Based at Tengah in Singapore it was joined by half of 64 Squadron in September 1964, however the remainder of the squadron back at Binbrook was gradually run down as the Tengah Flight was increased in strength to become a full squadron on 1 April 1965.

The increased Javelin presence in the Far East was the result of heightened tension in what became known as the Indonesia – Malaysia Confrontation. On 16 September 1963 Malaya and the British crown colonies of Sabah, Sarawak and Singapore were federated to become Malaysia, a process that had been opposed by the Indonesian government led by President Sukarno. With Indonesia threatening its newly formed neighbour, British forces in the area were reinforced and the Javelin took on a role for which it was never intended. As the last commander of 60 Squadron W/C Mike Miller describes some of the work that was carried out and the difficulties that were encountered

Our major operational commitment in the latter stage of 60 Squadron's life as a Javelin squadron was with the Indonesian Confrontation where we were operating on a very broad front over most of Malaysia with aircraft on standby at Butterworth and Tengah, and at Labuang and Kuching in Borneo. These were essentially readiness operations or, in the case of Borneo, patrols over the jungle which was pretty inhospitable country with very few aids and unpredictable weather with poor weather information. It was certainly quite a challenge for the crews. We usually operated as pairs so that one could keep an eye on the other and it was all low level flying so we did a lot of flying on one engine to conserve fuel and prolong the patrol time. However you had to be very careful to keep a good reserve of fuel to get back, especially towards the end of the day as big cumulo-nimbus clouds would build up and there were often heavy storms where you might have to divert. The work that we had to do was totally different from what the Javelin had been designed for but it coped very well indeed and of course it helped to have a navigator to share the load. For a time 60 and 20 (Hunter) Squadron operated together so that the pair that we flew consisted of one Javelin and one Hunter. At the height of Confront-ation in 1964–65 we were reinforced by 64 Squadron who flew Javelin FAW.9R's which had the capability of taking long-range fuel tanks under the wings. However they all operated as 60 Squadron over the huge area that it was our responsibility to look after and so the squadron (on paper) was the largest fighter squadron ever as it had something like twenty-four aircraft on strength.

For our practice interceptions we mainly operated against Hunters, Canberras and Vulcans on detachment but the Indonesians were dropping terrorists into the jungle from C-130 Hercules. To get a missile acquisition on a turbo-prop was slightly different as you were not looking at a jet efflux straight out of the back, rather one that was coming out sideways. We were most concerned however about the P-51 Mustang. Although this was a World War Two vintage single-engine fighter it was still very fast and was capable of speeds of over 400 mph. Operating from airfield's in Java it would have had a very short run time to Singapore and it was a major threat for us. We did a lot of low level interception practice out in the Singapore straits right down to 50 ft, Javelin to Javelin, but we never had a suitable piston engine target. We did ask Boscombe Down to look at this and they did some trials with a Harvard and the missile would acquire, but it was extremely difficult as you had to come in from the side to get the exhaust from one side or the other. I think we would have been hard pushed to get a Mustang at night if he just wanted to drop a bomb on Singapore as a nuisance exercise. For a guns attack there would obviously have been a good overtake capability but we would have needed a visual which would have been very difficult to achieve. Fortunately we were never challenged.

Perhaps the most difficult day for the Javelin force in the Far East occurred on 8 November 1965 when two of 64 Squadron's FAW.9R's were lost. The sequence of events started when XH887 flown by F/L Keith Fitchew, with F/L Evans as navigator, returned to base after a night PI and commenced a ground controlled approach. As he progressed through his pre-landing checks, F/L Fitchew found that he was unable to lower the undercarriage and the application of positive and negative 'g' during various manoeuvres, together with use of the emergency system made no difference at all. A visual check confirmed that the undercarriage was still fully retracted and as a wheels-up landing in a Javelin was not recommended, the decision was taken to carry out a controlled ejection over the sea at a height of 8,000 ft and a speed of 230 kts IAS. In the event the crew were able to eject successfully, although F/L Fitchew received back injuries.

As part of the rescue mission another of 64 Squadron's Javelin FAW.9Rs (XH959) was launched and this was flown by F/L Peter Poppe with F/L P.G.W. 'Buster' Unstead as navigator. The weather conditions at the time were clear with a visibility of thirteen miles with a full moon shining on a calm sea. After being homed onto the approximate position where the crew of XH887 had ejected, F/L Poppe descended to carry out a visual search. He levelled at a height of 500 ft before lowering the undercarriage and switching on the taxy lights to illuminate the surface of the sea. It is likely that he also lowered the flaps to the optimum setting for flying at low level and low speed. During the search a flare was sighted and the Javelin flew over a dinghy at a height of 'a couple of hundred feet'. A little later the aircraft flew into the sea and broke up. The Board of Inquiry that was set up to investigate the crash established that the navigator ejected, probably under water, and although he received spinal injuries, he survived. F/L Poppe also attempted to eject but he was drowned. The crash was witnessed by F/L Keith Fitchew from his dinghy and he was of the opinion that Peter Poppe had attempted a manual escape and was not strapped in when he activated his seat [Poppe had previously had to bale out of a Meteor T.7 and had also been forced to eject from a Meteor FR.9 which broke up in mid air].

Although flying over the jungle was extremely demanding for the crews, the very presence of a Javelin low overhead could be extremely effective and a technique was adopted to select reheat at very low level. This was used on a number of occasions to intimidate insurgents on the ground as the noise of the reheat cutting in was not dissimilar to the explosion of a bomb as W/C Mike Miller explains

Towards the end of February 1966 when the Confrontation was coming to an end 60 Squadron was accredited with one third of a group of terrorists. We had just arrived at Kuching on a deployment from Singapore when the Army called for a patrol in a certain jungle area where they suspected that a group of terrorists were hiding. A Javelin was hastily refuelled and took off to fly to the area in question. By shear good luck when flying at about 250 ft over the tops of the jungle, the navigator (S/L Geoff Moores) caught a suggestion of movement in a clearing out of the corner of his eye. The pilot (F/L Colin Holman) turned the aircraft round to beat up the clearing, diving and putting in the reheat which went in with a spectacular bang with lots of flame out of the back. He did this for as long as he could before fuel dictated that he should return. The Army were able to get there and captured six terrorists and on interrogation they were convinced that they were under attack and being bombed.

With much political instability throughout the area it was no great surprise when 60 Squadron were asked to fulfil another duty by operating out of Hong Kong. This was in response to the activities of communist agitators in Hong Kong and involved a lengthy transit flight from Singapore

When 64 Squadron disbanded in June 1967 we acquired their Javelin FAW.9R's as our own aircraft were ordinary FAW.9's. This gave us much greater capability and flexibility of operations as the new aircraft could carry long-range fuel tanks. Within a fortnight of taking over these aircraft the squadron was tasked suddenly with a special deployment to Hong Kong. There had been a Hunter squadron in the colony but it had been withdrawn and communist Chinese troubles started soon after. Javelins had never been to Hong Kong before and even with tanks on it was a challenging route which took us from Singapore to Labuan and then the American base at Clark Field in the Philippines and on to Kai Tak in Hong Kong. The most difficult leg was from Clark

Field to Kai Tak. Communications between ourselves and the Americans were not good; the weather information was not very reliable and the only diversion after the mid point would have been a Chinese airfield. As this was not a viable proposition there effectively wasn't a diversion and so at the mid point you had to decide whether to go back to Clark Field or carry on to Kai Tak. To enable us to make that decision a Canberra flew out from Kai Tak with the latest weather and a Shackleton was also laid on for low level air-sea rescue. There were a total of three detachments to Hong Kong between July 1967 and early 1968. We carried on with our regular programme of practice interceptions as well as performing naval liaison with the minesweepers that were based there. In addition we flew patrols over the sea looking for illegal junk operations. As these patrols were flown at very low level we had some frights when we were suddenly confronted with junks and other shipping

The Javelin was retired from RAF squadron service when 60 Squadron was disbanded at Tengah on 30 April 1968. To mark the occasion it was decided that the aircraft would go out in style with full use being made of the FAW.9's reheat system

Because of the infinitely variable control of the airbrakes on the Javelin it was perfectly possible to fly formation in reheat. The reheat itself was either in or out, you had no control over the power, but you could control the speed of the aircraft with the airbrakes. The trick was for the leader to select half airbrake then the box formation would formate using airbrakes rather than throttle. The disbandment parade was held at sunset and we did a flypast in diamond nine formation. After the inspection, we came back with a box four in complete darkness with nice long flames coming out of the back, something that I don't think had been done before. We only practised it once because it was the end of the time for the aeroplane. We found that when the aircraft broke to go round the circuit downwind the pilots of the formating Javelins had destroyed their night vision and they couldn't see each other. To get round this we mounted Canberra downward ident lights between the engines on the back panel of the Javelin fuselage, and when we looked at them after the flypast, the lights of all four aircraft were on the point of meltdown!

With the disbandment of 60 Squadron the Javelin left RAF squadron service after twelve years in which its role had changed dramatically from being the UK's principal means of defence against bombers armed with nuclear weapons to its use in the Far East where, in addition to air defence duties, it was also used widely for low level patrols and as a psychological weapon in guerrilla warfare. A single Javelin was to survive longer than all the others however as FAW.9 XH897 continued to fly with A&AEE at Boscombe Down until early 1975. During this time it was used as a photo chase aircraft and it was also used in the Concorde and MRCA test programmes for calibration duties. Its last flight was on 24 January 1975 to Duxford where it remains as part of the Imperial War Museum collection.

Glossary

A&AEE	Aeroplane and Armament Experimental Establishment
AGL	Above Ground Level
AI	Airborne Interception (radar)
AIB	Accident Investigation Branch
AMSL	Above Measured Sea Level
AOC	Air Officer Commanding
AOG	Aircraft on Ground
ASP	Aircraft Servicing Platform
ATAF	Allied Tactical Air Force
ATC	Air Traffic Control
AWDS	All-Weather Development Squadron
AVTAG	Aviation Turbine Gasoline
AVTUR	Aviation Turbine fuel
BAC	British Aircraft Corporation
CA	Controller (Air)
CFE	Central Fighter Establishment
CFI	Chief Flying Instructor
CFS	Central Flying School
CRT	Cathode Ray Tube
DME	Distance Measuring Equipment
DV	Direct Vision panel

GLOSSARY

ECM	Electronic Counter Measures
FAW	Fighter All-Weather
FCIRS	Fighter Command Instrument Rating Squadron
FIS	Fighter Identification System
FL	Flight Level
FTS	Flying Training School
FTU	Ferry Training Unit
GCA	Ground Controlled Approach
GCI	Ground Controlled Interception
Gee	Radio aid to navigation
GGS	Gyro Gunsight
HGU	Horizon Gyro Unit
HP	High Pressure (fuel cock)
IAS	Indicated Airspeed
IFF	Identification Friend or Foe
ILS	Instrument Landing System
IMN	Indicated Mach number
JMTU	Javelin Mobile Training Unit
JPT	Jet Pipe Temperature
kts	Knots
MPC	Missile Practice Camp
MRCA	Multi-Role Combat Aircraft
MU	Maintenance Unit
NATO	North Atlantic Treaty Organisation
OC	Officer Commanding
OCU	Operational Conversion Unit
ORB	Operations Record Book
ORP	Operational Readiness Platform
PAR	Precision Approach Radar
PI	Practice Interception
PPI	Plan Position Indicator
PSI	Pounds per Square Inch
'q' feel	Artificial feel system fitted to power controls designed to give the pilot control feel proportional to aircraft speed
QFI	Qualified Flying Instructor

QGH	A ground controlled letdown procedure
QRA	Quick Reaction Alert
RAE	Royal Aircraft Establishment
RCAF	Royal Canadian Air Force
Rebecca	A navigation instrument that gave homing and range information
rpm	Revolutions per Minute
R/T	Radio Telephony
SBAC	Society of British Aircraft Constructors
SHAPE	Supreme Headquarters Allied Powers Europe
TMN	True Mach Number
VHF	Very High Frequency
UAS	University Air Squadron
UHF	Ultra High Frequency
USAF	United States Air Force

Index

INDEX

Smith, Brian 129
Smith F/L B. 137–8
Smith, W/C Dave 168
Smith, W/C Peter 193
Spurr, F/L Bruce 163–4
Stark, F/O F.H.B. 113
Sukarno, President 197
Sykes, F/L Gordon 178, 180–1

Thorn, F/L Pete 183–9
Thorn, S.A. 'Bill' 184
Thorne, Len 7

Unstead, F/L P.G.W. 200

Wark, F/O R.J. 140
Warner, F/O Tony 24, 150
Waterton, S/L W.A. 3, 6–10
Watson, F/Sgt Doc 149
White, W/C Harry 24, 26, 28, 49
Wilson, F/L J.S. 130
Wilson, S/L Tug 190
Wood, F/O E. 130
Woods, F/O Guy 156–61
Wright, F/L Arthur 155
Wright, W/C J.A. 132, 136
Wyborn, F/L D.J. 129

Bases/Locations
Aberporth 83, 166, 174
Aden 185, 187
Akrotiri 130, 134, 141, 160,
    189–91
Ashey 69
Aston Down 149

Bahrain 141, 170, 187
Baginton 73
Balsham 142
Bangkok 170

Bassingbourn 135
Benares 170–1
Benson 148
Bentham 5
Binbrook 145–6, 181, 197
Boscombe Down 7, 11–12, 69, 100,
    116, 161, 199, 203
Bournemouth 116
Brize Norton 7
Bruggen 117
Butterworth 170, 177, 198

Calcutta 170–1
Changi 171, 180
Church Fenton 25
Clark Field (Philippines) 201–2
Coltishall 49, 113, 139, 141, 162,
    175, 177, 184–5
Coningsby 161
Cranwell 160, 188, 190

Decimomannu 143
Delhi 170
Dezful Vahdati 187
Dingden 115
Diyarbakir 170, 187
Duxford 132–4, 136, 140–1, 203

El Adem 135–6, 143, 170, 185

Farnborough 9–10
Filton 67

Geilenkirchen 72, 110, 118, 121,
    127, 167–8, 197
Gutersloh 168, 188

Hartlepool 130
Hatfield 137
Headley Court 181, 192

209